T0205584

# Springer Theses

Recognizing Outstanding Ph.D. Research

## Aims and Scope

The series "Springer Theses" brings together a selection of the very best Ph.D. theses from around the world and across the physical sciences. Nominated and endorsed by two recognized specialists, each published volume has been selected for its scientific excellence and the high impact of its contents for the pertinent field of research. For greater accessibility to non-specialists, the published versions include an extended introduction, as well as a foreword by the student's supervisor explaining the special relevance of the work for the field. As a whole, the series will provide a valuable resource both for newcomers to the research fields described, and for other scientists seeking detailed background information on special questions. Finally, it provides an accredited documentation of the valuable contributions made by today's younger generation of scientists.

## Theses are accepted into the series by invited nomination only and must fulfill all of the following criteria

- They must be written in good English.
- The topic should fall within the confines of Chemistry, Physics, Earth Sciences, Engineering and related interdisciplinary fields such as Materials, Nanoscience, Chemical Engineering, Complex Systems and Biophysics.
- The work reported in the thesis must represent a significant scientific advance.
- If the thesis includes previously published material, permission to reproduce this must be gained from the respective copyright holder.
- They must have been examined and passed during the 12 months prior to nomination.
- Each thesis should include a foreword by the supervisor outlining the significance of its content.
- The theses should have a clearly defined structure including an introduction accessible to scientists not expert in that particular field.

More information about this series at http://www.springer.com/series/8790

Piotr Antonik

# Application of FPGA to Real-Time Machine Learning

## Hardware Reservoir Computers and Software Image Processing

Doctoral Thesis accepted by
the Université libre de Bruxelles, Brussels, Belgium

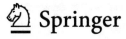 Springer

*Author*
Dr. Piotr Antonik
CentraleSupélec
Metz
France

*Supervisor*
Prof. Serge Massar
Université libre de Bruxelles
Brussels
Belgium

ISSN 2190-5053          ISSN 2190-5061   (electronic)
Springer Theses
ISBN 978-3-030-08164-5          ISBN 978-3-319-91053-6   (eBook)
https://doi.org/10.1007/978-3-319-91053-6

Printed on acid-free paper

This Springer imprint is published by the registered company Springer International Publishing AG part of Springer Nature
The registered company address is: Gewerbestrasse 11, 6330 Cham, Switzerland

*To my Dad*

# Supervisor's Foreword

Today, we have two completely different examples of computing systems. The first is our digital electronic computers that represent information using binary signals, 0 and 1, and process the information sequentially at a very high rate (GHz). The second is the biological brain that represents information in an analogue way (time and intensity of spikes) that processes the information in parallel, but rather slowly (kHz).

We do not really understand how the brain processes information. But the booming field of artificial intelligence and artificial neural networks tries to emulate the brain (if not in detail, at least imitating in some aspects how it processes information) using digital computers.

An alternative approach is to develop analogue hardware information processing systems that imitate, again only in some aspects, how the brain processes information. This line of research has a long history. It is closely related to advances in machine learning, as novel algorithms in machine learning can lead to novel hardware architectures.

In parallel with the above, photonics has developed into a major industry, which underlies many aspects of our information-driven society. But it remains extremely challenging to process information in the optical domain, and one is almost always obliged to carry out costly optical to electronic conversion whenever optical information must be processed (e.g. for routing).

A recent development in machine learning is the algorithm known as 'reservoir computing'. This algorithm turns out to be very well suited to experimental implementation, as there is a lot of flexibility how exactly it is implemented. This led to a number of photonic implementations that go beyond what had been achieved previously in optical computing. Photonic reservoir computing is a growing research field. The most important question in the area is whether these systems can be brought to the level where they can compete with existing solutions for optical information processing.

In his thesis, Piotr Antonik pushed the performance of these photonic reservoir computers considerably beyond what had been realised before, thereby expanding their capabilities and the range of possible applications. This was achieved by interfacing a photonic reservoir computer with a high-speed electronic device, a Field-Programmable Gate Array (FPGA), enabling interactions with the reservoir computer in real time.

The main results obtained during the thesis are:

(1) The experimental demonstration that photonic reservoir computers can solve tasks that change in time, which is important for some applications, particularly in telecommunications. For this, Piotr had to adapt the method known as online training to photonic reservoir computers.
(2) The experimental demonstration of a reservoir computing system whose hardware is used both for computing and for optimising its internal parameters, thereby considerably improving performance. By carrying out the optimisation in hardware, this work demonstrated that one could potentially circumvent the difficulties of previous work that relied on detailed numerical models of the experiment to realise the optimisation.
(3) The experimental demonstration that photonic reservoir computers with output feedback could produce periodic time series and emulate chaotic dynamical systems. This opens up a whole new area of applications of analogue reservoir computers, as well as novel conceptual questions about chaotic dynamics.
(4) The numerical demonstration (which Piotr Antonik hopes to soon confirm in hardware) that the performance and robustness of the output layer of reservoir computers could be considerably improved by online training.
(5) Real-time image analysis of optical coherence tomography of atherosclerotic arteries, using methods from artificial intelligence and FPGA programming. (In this chapter, Piotr Antonik used his expertise in machine learning and FPGA programming to make progress on a very different problem, namely real-time image analysis of optical coherence tomography of atherosclerotic arteries).

The thesis may give rise to industrial valorisation. First, reservoir computing is simple, flexible, easy to train and can be implemented in optics. It is believed that the most promising area of application in the short term is telecommunication through optical fibres, and in particular equalisation of optical communication channels (i.e. corrections of the distortions that occur during transmission). Second, Piotrs work on optical coherence tomography may give rise to applications on a short timescale. Indeed real-time image analysis of optical coherence tomography of atherosclerotic arteries may help expand the use of this technique in hospitals.

Piotr Antoniks thesis is highly interdisciplinary. The theory on how to design, train, exploit and benchmark the experiments was supplied by ideas from machine learning, and in particular from the topic of reservoir computing. The experiments themselves were built from photonic and electronic components. Understanding how they would perform required extensive numerical simulations. To develop new features, Piotr used an advanced tool from the electronics world—a FPGA chip. Programming of such a chip is a complex task that requires in-depth expertise of analogue and digital electronics.

Photonic reservoir computing is a rapidly developing area. Piotr Antoniks thesis is very clearly written. It gives a snapshot of some of the latest advances in the area. I hope that its publication in the 'Springer Theses' series will help spread knowledge and interest about this area.

Brussels, Belgium                                                                    Prof. Serge Massar
March 2018

# Abstract

Computers have gained a cardinal place in modern societies, thanks to higher efficiencies and miniaturisation. However, their dramatic progress will soon have to stop as the limits of miniaturisation are being reached. Furthermore, few people realise that those computers are, in fact, not as powerful as they seem to be. And while the world champion at Go lost to a computer, an average human still beats a computer at relatively easy tasks such as recognising an object in a picture. Artificial intelligence is the key to more versatile computing machines capable of solving such challenging tasks.

My Ph.D. research lies in the intersection of artificial intelligence—the science of solving complex problems in a smart way, and photonics—the science of light that offers the potential to design ultra-fast and energy efficient processing devices. Photonics, and in particular optical telecommunications, underlies much of our information society, allowing information transmission at unprecedented rates through optical fibres. Recent optical realisations of simple artificial neural networks, also known as reservoir computers, have allowed a breakthrough in optical computing. I have pushed the performance of these systems significantly beyond what was achieved before. By interfacing a photonic reservoir computer with a high-speed electronic device, a FPGA, I could interact with the reservoir computer in real time, and thereby considerably expand its capabilities, and therefore the range of possible applications.

Specifically, we demonstrated experimentally that (1) photonic reservoir computers could solve tasks that change in time, which is important for some applications, particularly in telecommunications, (2) reservoir computing systems could use their hardware both for computation and for optimising its internal parameters, thereby considerably improving their performance, (3) photonic reservoir computers with output feedback could produce periodic time series and truthfully emulate chaotic dynamical systems. Furthermore, we presented a numerical demonstration (which we soon hope to confirm experimentally) that the performance and robustness of the output layer of reservoir computers could be considerably improved by online training.

Finally, I used my expertise in machine learning and FPGA programming to make progress on a very different problem, namely real-time image analysis of optical coherence tomography of atherosclerotic arteries.

# Preface

This dissertation contains the full story of the 4 years of my Ph.D. The structure of the document is quite simple. The first chapter explains the theoretical and experimental basics. Throughout my thesis, I worked on four distinctive experiments. Some of them were my own projects, others were collaborations with other researchers in our group. They will be described in four separate chapters (Chaps. 2–5). By the end of my Ph.D., I took a 5-month internship at the University of Texas at Austin, that will be covered in Chap. 6. Finally, Chap. 7 concludes the story with a few ideas for future research.

Before writing this thesis, I had to make a crucial choice: either spend 3 to 5 months writing an original dissertation from scratch, or fill the thesis with my publications and spend the remaining time on another experiment. Without much hesitation, I chose the second option. In other words, this thesis compiles the work presented in these journal papers, properly integrating them to form a continuous story. This is a somewhat lazy approach—I do not deny it. But I believe the importance of a thesis consists of its scientific value. And those extra 5 months allowed me to complete another interesting project, thus increasing the significance of my work.

Another word of warning should be written concerning the style of the present thesis. Scientific English is a very clear and concise communication tool, but may seem somewhat boring. And after writing a few journal papers and a dozen of conference proceedings, I wanted to add some colours to the final publication of my Ph.D. Therefore, while its tone remains scientific most of the time, I allowed myself a few minor digressions. The reader will notice that from the very first lines of the first chapter.

Final remark, most chapters contain a 'bonus' section, describing the challenges encountered during the realisation of a particular project. These sections contain the back story of each experiment. In most fields of science, positive results are published, and the negative remain in the shadows. However, knowing what has been tried but did not work may save time in some cases, or even inspire new ideas. For this reason, I decided to include in this dissertation some facts that did not make it to the journal papers.

Brussels, Belgium                                              Piotr Antonik

# Acknowledgements

It is an immense pleasure to thank the numerous people who made this thesis possible.

First and foremost, it is difficult to overstate my gratitude to my Ph.D. supervisor, Prof. Serge Massar. There are two things I long for the most—freedom and support—and Serge gave me both. While freedom gives birth to that spark that ignites new ideas, support is the fuel that turns ideas into projects and, ultimately, results and publications. So thank you, Serge, for being the best supervisor I could ever wish for. This paragraph would be dramatically incomplete without a big thanks to Prof. Marc Haelterman, my co-supervisor, for his everlasting support.

The next hat tip goes to the team of awesome postdocs I had a great pleasure of working with. I could never complete this thesis alone, and these are the people who took direct part in some of my projects. Starting with Dr. François Duport, who took me under his wing right from the start, before I even started as a Ph.D. student, and taught me everything I needed to work independently in the lab, and even more. Greeting François in the office as early as 7 a.m. and regularly seeing him in the lab after 9 p.m. constantly reminded me that the only time success comes before work is in the dictionary.[1] Following with Dr. Anteo Smerieri, who 'basically' guided me through the complex theory of reservoir computing and learning methods. Few people can speak of intricate algorithms with simple words, and even less could turn it into an enjoyable show, with particularly well-placed jokes—as was demonstrated on numerous occasions. Last (chronologically) but not least, comes Dr. Michiel Hermans, from whom I acquired a much deeper understanding of the machine learning field in general, and reservoir computing in particular. And on top of that, the idea of taking an internship abroad was inspired by Michiel, which ultimately led to an amazing experience in Texas (more on that in Chap. 6). Thank you very much for that!

I am particularly grateful to Prof. Thomas Milner for offering me the life-changing opportunity to join his research team at the University of Texas.

---

[1] Quote credited to Stubby Currence by Quote Investigator.

Working with Prof. Milner was a very inspiring experience, and I appreciate how much I could learn in so little time about various aspects of scientific life.

I would like to thank all the Jury members for their valuable comments on the present thesis and, in particular, Dr. Daniel Brunner for his in-depth proofreading of the manuscript and the long list of questions and comments, that made this work more accurate and complete.

Debuting in electronics, and especially in FPGA design, is all but an easy task. Fortunately, I could benefit from valuable help from several people well skilled in this art. Thus, I would like to express my very great appreciation to Benoit Denègre for providing a solid starting point, as well as Colin Fera, Matthew Luscher, Ashkan Ashrafi and Arnaud from 4DSP for providing precious documentation and technical support.

The working environment is only as good as the people who surround you. In that sense, OPERA-Photonique is the second best thing that happened to me on this journey. Never before could I imagine that scientific research could be accompanied by numerous fun parties, uncountable pies, video games and movie nights. Therefore, an enormous shout-out goes to my co-workers, in alphabetical order: Akram Akrout, Marc Bauduin, Serena Bolis, Arno Bouwens, Thomas Bury, Ali Chichebor, Charles Ciret, Robin De Gernier, Rima Dadoune, Evangelia Dimitriadou, Evdokia Dremetsika, Michael Fita Codina, Simon-Pierre Gorza, Wosen Kassa, Pascal Kockaert, Virginie Lecocq, François Leo, Anh-Dung Nguyen, Laurent Olislager, Nicolas Poulvellarie, Maïté Swaelens, Guillaume Tillema and Quentin Vinckier. Very special thanks to Prof. Philippe Emplit for creating and maintaining such a productive environment, and to our awesome secretaries, Ibtissame Malouli and Alexandra Peereboom, who just took care of everything.

And when I could not stand the guys from OPERA anymore (just kidding), I could always join my always welcoming colleagues at LIQ, again in alphabetical order: Cédric Bamps, Olmo Nieto Sileras, Jonathan Silman, Tom Van Himbeeck and Erik Woodhead. And again, warm thanks to the secretaries, Sabrina Serrano Alvarez and David Houssart, for handling my orders, travel documents, and, most importantly, reimbursing me for all my expensive conference trips.

The one thing I benefited the most from scientific conferences—besides the chance to travel to some exotic places on the globe—is the opportunity to interact with scholars and industry experts, absorbing as much knowledge from them as possible. My first ICONIP conference in Istanbul, and long discussions with Prof. João Paulo P. Gomes were a particular revelation. I wish to acknowledge here their valuable help.

Most of the first chapter of this thesis, as well as one or two papers were written in various hospitals, clinics and medical centres. I very much appreciate the efforts of the staff who made the task of writing in waiting halls quite a comfortable exercise.

I have composed quite a long list so far, but still I have the impression that I missed someone. To those people I offer my sincere apologies—for my poor memory, and my gratitude—for their valuable help.

I would like to conclude with people who made a rather indirect contribution to this work. My big thanks go to my school teachers for letting me do what I really wanted (that is, solving mathematical, and later, physical problems) and not paying much attention in class, my school and university buddies for helping me get through the education process with that much fun, my close friends Nicolas De Groote, Livio Filippin, Jonathan Bloch and Anton Leonov for their support and inspiration. And of course, many sweet thanks go to my dear Luda, and her sister Sveta, for taking such good care of my new hairstyle.

Most appreciation and gratitude usually goes to people for their positive contributions. However, as an old saying goes—there can be no evil without good—I would like to make an exception and express my gratitude to all people who managed to hurt me, deeply or not, intentionally or not. Thank you for making me that much stronger!

Family is a true masterpiece of nature, and undoubtedly the most precious treasure one gets to cherish. And while most of my family lives abroad and quite far away, I felt a very positive lift every time I went to visit them. Thank you for filling me with confidence, love and kindness! And what a person would I be if I failed to mention my beloved sister Maria for her artistic touch and a very special character. You rock!

And as the best is usually saved for the end, my warmest thanks go to my parents. This is where words begin to fall short, but I will try my best anyway. To my mum, thank you for being such a positive, dynamic, kind, caring and forgiving person. Few grown-ups think twice before leaving their parent's home these days, but somehow, you made me think and rethink a gazillion times. And to my dad, thank you for being a model to me for so many years, a friend and guide I could follow anytime with my eyes closed. Thank you for introducing me to arithmetic and basic algebra at the age of six, and for directing me straight into my current scientific life. There are few things one has the luxury of being certain of. But for me, not for a second did I doubt that one day I would be here, preparing the defence of my Ph.D. thesis. And although I can no longer learn from you as I used to, I can still follow that bright star in the night sky you turned into.

# Contents

# Chapter 1
# Introduction

In this chapter we will address three questions: (1) What is reservoir computing? (2) What does it have to do with optics and electronics? (3) What are FPGAs? That is a lot of information to cover, so let us get started right away!

## 1.1 From Machine Learning to Reservoir Computing

Reservoir computing—what a peculiar concept! Are we talking about a bucket of water performing computations? The idea may seem weird, but…it is actually not far from reality! In fact, there has been an experiment carried out in a water tank, where ripples on the surface of water were sampled and used to process information [1]. But this is not exactly what reservoir computing is all about. Attributed to the machine learning (ML) field—a subfield of computer science [2–6] that studies data processing algorithms capable of learning from the data itself—reservoir computing is not an algorithm *per se*, but rather a set of ideas that significantly simplify another algorithm and make it more suitable for practical applications. This other algorithm, or, rather, a class of algorithms, is called artificial neural networks. To understand the whole story, we need a general overview of the said machine learning field.[1] The goal of this section is thus to present to the reader the bigger picture, following a top-down approach. We will start with an overview of machine learning, with some basic ideas and several examples. Then, we will dive into artificial neural networks, again leaving aside most of unnecessary technical details. Finally, within neural networks we will finally introduce the RC paradigm, now with all mathematical details needed to understand how it works.

---

[1]This is obviously a debatable point. But it did work for me—my true revelation on reservoir computing, how and why it works, happened when I saw what is around—so I am going to stick to this plan.

© Springer International Publishing AG, part of Springer Nature 2018
P. Antonik, *Application of FPGA to Real-Time Machine Learning*,
Springer Theses, https://doi.org/10.1007/978-3-319-91053-6_1

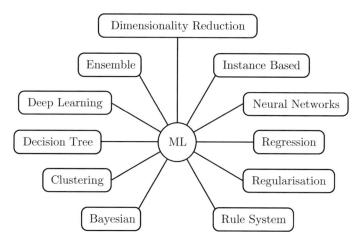

**Fig. 1.1** Map of the machine learning field. Far from being the most exhaustive, it is sufficient to show what algorithms, or classes of algorithms can be found out there. Figure inspired by the Mindmap from Machine Learning Mastery

### 1.1.1 Machine Learning Algorithms

ML enjoys a fast evolution in these days, as people are desperately looking for methods to efficiently process the huge amounts of data coming from everywhere, and ML offers several very promising solutions [7–12]. Figure 1.1 draws a more or less complete picture of the machine learning field. Here we will overview a few of these methods (the most popular ones) with their basic properties and applications, obviously simplifying the details to the bare minimum. The goal here is not to review the machine learning field, but to give the reader a broad view of the algorithms that can be found there.

**Decision trees**:   Commonly used in statistics and data mining, decision trees are predictive models for data classification based on its properties [7, 9, 13, 14]. In a simple decision tree, the leaves are labelled with all possible classes. On its way from the root to leaves, the input instances "travel" through decision nodes (where branches of the tree split), where data parameters define the following path.

**Bayesian networks**:   A Bayesian network is a probabilistic graphical modelling technique used in computational biology, bioinformatics, medicine, engineering, and many other domains [10, 15, 16]. A directed acyclic graph represents the data as a set of variables and their conditional dependencies, which allows to draw probabilistic relationships between data features.

**k-nearest neighbour**:   Instance-based algorithms, such as k-nearest neighbour [17–19], typically build a database of examples and compare the incoming data using a certain similarity metric in order to find the best match and make a prediction. They are often used for dimension reduction, i.e. removing unnecessary redundancies from very large sets of data.

**Support vector machines**: Commonly employed as linear classifiers for e.g. text or image processing, SVMs [20–23] map the input data into a high-dimensional space, using specific algorithms, where different classes can be separated (clustered) by a set of hyperplanes.

**Artificial neural networks**: Family of models, inspired by biological neural networks, used to estimate or approximate (generally unknown) functions depending on a large number of inputs [24–27]. They come in different shapes and flavours, and besides data processing, they are also used in neuroscience.

**Deep learning**: A class of ML algorithms that cascade multiple information processing layers, each successive layer receiving the output of the previous one as input [28–31]. The layers learn multiple levels of data representation, that correspond to different levels of abstraction, and form together an hierarchy of concepts. The most successful deep learning methods involve neural networks and have shown breathtaking results in speech and image recognition, natural language processing, drug discovery and recommendation systems. Other less known deep architectures exist, such as multilayer kernel machines.

To process data, these algorithms need to be trained—in other words, taught what to do with the data. Remember, ML algorithms are not designed to perform well on a particular dataset, but rather to execute a certain versatile task. The training serves to fine-tune the algorithm for better performance on the dataset of interest. The training can be done using various techniques, commonly grouped into categories, based on their action principle.

**Supervised learning**: The algorithm is presented with a labelled dataset, that is, where the output is known for each input, such as spam/not-spam classification or a set of tagged images [32, 33]. During the training process, the model is tuned to correctly classify all the inputs, and then tested on a new set of data, that was not used for training. This process is carried on until a desired level of accuracy is achieved on the test set.

**Reinforcement learning**: Inspired by behavioural psychology, this methods is employed when the corrects outputs or labels are unavailable [34, 35]. Instead, the algorithm is supplied with a reward (or error) function and then optimised to maximise (or minimise) it. Such approach is commonly used in robotics, where exact movement patterns of different motors or actuators are unknown, and the robot is trained to optimise the reward function, given by e.g. the distance travelled.

**Unsupervised learning**: As the name suggests, here the algorithm does not use any labelled dataset nor reward function [5, 36–38]. It is presented with the data alone and is supposed to find an underlying structure or some hidden insights. This case is the hardest to understand, as it looks like some kind of dark magic. Since I have never used such methods, we shall leave the details aside. A typical example of unsupervised learning is clustering, that is, the task of grouping a set of objects by similarity.

Other approaches exist, such as semi-supervised learning [39], but they lie beyond the scope of this introductory overview.

To sum up this section, numerous machine learning algorithms exist, based on various approaches and suited for different tasks. To process data, they need to be trained first, and this can also be done in various ways, depending on the task and the type of data available. Among all the methods lies the family of artificial neural networks. And since reservoir computing has something to do with neural networks, let us discuss them in detail in the next section.

## 1.1.2 Artificial Neural Networks

The first model of artificial neural networks (ANN), introduced in 1943 [40] split the research in two distinct approaches: the study of actual biological processes in the brain on one side, and application of neural networks to machine learning. The research stagnated after the discovery of a fatal flaw: basic neural networks (also known as perceptrons—we will introduce them very soon) were incapable of processing the basic exclusive-or (XOR) circuit! [41]. On top of that computers did not have enough power to handle large networks on the long run. Later on, the CMOS technology (that lead to an explosion of computational speed) and the novel backpropagation algorithm [42, 43] allowed to efficiently train large multi-layer networks. Recent advances in GPU-based implementations and the emergence of highly complex, deep neural networks made this approach very popular and brought breathtaking results in e.g. speech or text recognition and novel drug discovery.

Let us take a look inside those networks. They are composed of elementary computation units—neurons. A biological neuron is a cell capable of producing a rapid train of electric spikes. Its complex internal dynamics can be described by the well-known Hodgkin-Huxley model [44] that takes into account the exact three-dimensional morphology of the cell. Simulating such a precise model is extremely demanding in computational power, and so is, although of great interest for brain research, impractical for real-world applications. For this reason, artificial neurons have been introduced, keeping the spiking behaviour but greatly simplifying the internal dynamics. A plethora of models have been proposed to emulate artificial neurons (see e.g. [45–48]). All of them encode information into spike trains, just as we think biological neurons do. But one can simplify the neuron one step further and remove the spikes at all by defining the average spiking frequency $a$. Such neurons are called analogue neurons and their behaviour is described by the following simple equation

$$a = f\left(\sum w_i s_i\right), \tag{1.1}$$

where $a$ is the output of the neuron (that can also be referred to as the current state of the neuron, or the activation), $s_i$ are the inputs coming from the neighbour neurons in the network, $w_i$ are the weights of these connections (thus making it possible to create weak or strong connections between neurons), and $f$ is the activation function, that describes how the neuron reacts to its inputs. Crucially, this simplification removes

Input layer                        Hidden layer                        Output layer

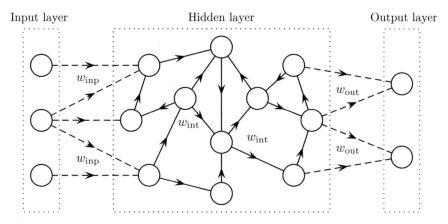

**Fig. 1.2** Example architecture of an artificial neural network. The neurons are grouped in three layers—input, hidden and output—based on their connections with the outside world. The network may contain several hidden layers (this example has only one)

the complex temporal dynamics of the neurons and make discrete-time computations possible. This, in turn, allows to simulate large numbers of neurons with relatively low computational power.

The neurons are gathered in network-like structures with three main characteristics.

**Architecture**:  It defines the size of the network and the connections between the nodes, which in turn defines how they exchange information. An example neural network is sketched in Fig. 1.2. The circles denote the nodes, or the neurons, and the arrows show the connections from the output of a neuron to the input of another. The neurons are commonly categorised into three layers, based on their role in the network. The input layer nodes receive signals from outside and output layer neurons produce output signals of the network. The other neurons, as they cannot be accessed from the outside of the network, are called hidden neurons, and can be grouped into one or several layers. All connections, depicted with arrows, are parametrised with associated weights—input, output or internal—which define the strength of the connections.

**Activation function**:  The activation function defines the individual behaviour of the neurons, that is, how they respond to input signals. To avoid unconstrained dynamics of the network, the activation function should be bounded, usually within $[-1, 1]$. The sigmoid function is one of the most popular choices, alongside the so-called linear rectifier function [49]. Other functions, such as hyperbolic tangent or sine, are also used.

**Tunable weights**:  Artificial neural networks are valued for their ability to learn by means of adjusting their weighted connections (input, output or internal). Under supervised learning paradigm, for instance, the network is fed with numerous input instances, and the output is compared to the desired output. Various training

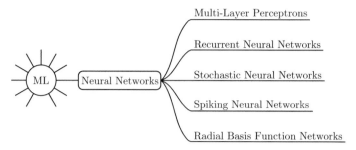

**Fig. 1.3** Several examples of neural networks

algorithms can then be used to adjust the weights so that the network output signal matches as closely as possible to the target output.

Artificial neural networks come in many different shapes and flavours. We will limit this introduction to a few notable examples, shown in Fig. 1.3, leaving the complete list to specialised literature [49].

**Multi-layer perceptron**:    A MLP is a feedforward artificial neural network [27, 36, 50]. That is, the information flows in one direction, from input to output neurons (through the hidden ones) with no cycles or loops in the network.[2] Owing to a nonlinear activation function, MLPs are capable of partitioning data that is not linearly separable. They found many applications in speech or image recognition in the 1980s, but have been superseded by much simpler support vector machines (see Sect. 1.1.1) in the 1990s.

**Recurrent neural network**:    Unlike feedforward networks, RNNs are allowed to form directed cycles between neurons, which allows them to exhibit temporal behaviour and adds internal memory [31, 51, 52]. That is, the network can "remember" the previous inputs and its current state is no longer entirely defined by the current input. This makes them a powerful tool that can be applied to digital signal processing, speech and handwriting recognition.

**Stochastic neural network**:    Stochastic networks are built by introducing randomness into the system, either by means of a stochastic transfer function, or by assigning random weights [53, 54]. This makes them suitable for optimisation tasks, as local minima are avoided with these random fluctuations. They have found applications in e.g. bioinformatics and drug discovery.

**Spiking neural networks**:    Spiking neurons increase the level of realism by incorporating the temporal dynamics in their operating principle [55–58]. Similarly to biological neurons, spiking neurons do not produce an output at each update cycle, but rather fire a spike whenever their internal states reaches a certain threshold. They have been used in studies of biological neural circuits, since they can model simple central nervous systems. However, because of the increased computa-

---

[2]Note that the example in Fig. 1.2 does contain several loops.

tional power required to simulate these realistic networks, they are yet to find useful applications in engineering.

**Radial basis function networks**: A radial basis function is a real-valued function whose values only depend on the distance from the origin. Neural networks, based on these functions, are composed of an input layer, one hidden layer with nonlinear radial basis activation function neurons and a linear output layer [59–61]. Such structures can, in principle, interpolate any continuous function and have been shown to be more advantageous on complex pattern classification problems. Mathematical proofs and further details can be found in [49].

This concludes our brief overview of machine learning and artificial neural networks. Let me say again that the purpose of this introduction was not to turn the reader into expert in machine learning, but merely show the general context of this work. In the next section we will focus on the main topic of interest—reservoir computing—with much more in-depth discussions.

### 1.1.3 Reservoir Computing

Reservoir Computing (RC) is a set of machine learning methods for designing and training artificial neural networks, introduced independently in [62] and in [63]. The idea behind these techniques is that one can exploit the dynamics of a recurrent nonlinear network to process time series without training the network itself, but simply adding a general linear readout layer and only training the latter. This results in a system that is significantly easier to train (since one only needs to optimise the readout weights), yet powerful enough to match other algorithms on a series of benchmark tasks.

These ideas can be applied to both recurrent and spiking recurrent neural networks, which gave birth to two concepts called Echo State Networks (ESN) [64] and Liquid State Machines (LSM) [63], that are grouped under the reservoir computing paradigm. An ESN is a sparsely connected, fixed RNN with random input and internal connections. The neurons of the hidden layer, commonly referred to as the reservoir, exhibit nonlinear response to the input signal due to a nonlinear activation function (hyperbolic tangent seems to be the most common choice). Liquid state machines rely on the same concept, but the reservoir consists of a "soup" of spiking neurons. The name "liquid" comes from an analogy to ripples on the surface of a liquid created by a falling object. Interestingly, this concept has actually been implemented in hardware, that is, as the name suggests…in a tank full of water! [1].

For hardware reasons, as will become clear in Sect. 1.2, in this work we will only deal with analogue neurons, leaving the spiking models aside. From now on, to simplify the ideas, I will make no distinction between Echo State Networks and Reservoir Computing.

It is now time to introduce the math used describe the dynamics of a reservoir computer. Let us denote the neurons (also called nodes, or internal variables of the reservoir) $x_i$. As they are analogue neurons (see Sect. 1.1.2), we may consider that they evolve in discrete time $n \in \mathbb{Z}$, so we note them $x_i(n)$. The index $i$ goes from 0 to $N - 1$, with $N$ being the reservoir size, or the number of neurons in the network. To fix the ideas, let us consider $N = 50$, since this is a value commonly used in experiments. Remember Eq. 1.1 giving the output of an analogue neuron? The evolution equation of a reservoir node is fairly similar and given by

$$x_i(n + 1) = f \left( \sum_{j=0}^{N-1} a_{ij} x_j(n) + b_i u(n) \right), \qquad (1.2)$$

where $f$ remains the nonlinear activation function, $u(n)$ is the external input signal that is injected into the system, and $a_{ij}$ and $b_i$ are time-independent coefficients that determine the dynamics of the reservoir. Specifically, $a_{ij}$ is called the interconnection matrix, since it defines the strengths of connections between all the neurons within the reservoir, with 1 being the strongest connection, and 0 meaning no connection. The vector $b_i$ contains the input weights and defines how strong is the input to each neuron. These coefficients are usually drawn from a random distribution with zero mean. As an alternative point of view, this equation can be expressed as follows

$$\begin{matrix} \text{Future state of} \\ \text{the i-th neuron} \end{matrix} = \begin{matrix} \text{Nonlinear} \\ \text{function of} \end{matrix} \left( \begin{matrix} \text{Previous \quad states \quad of} \\ \text{connected neurons} \end{matrix} + \begin{matrix} \text{Weighted} \\ \text{input signal} \end{matrix} \right).$$

This form emphasises the two major contributions to the reservoir dynamics: the feedback, that is, the previous values of the neighbour neurons and the input signal. This feedback is the recurrent part of the neural network that gives it internal memory, essential for some tasks (as will be discussed later in Sect. 1.1.4).

The concept of an Echo State Network suggests that (a) the connections between the neurons, given by the matrix $a_{ij}$ should be sparse (that is, a relatively low number of connections should be present within the network) and (b) the exact topology (or connection pattern) does not really matter. This is a considerable loss from the point of view of general RNNs, as all these connections "that do not matter" could be trained instead to better fine-tune the network. But from the point of view of ESNs, and especially their hardware implementations, this is a massive relief. It allows one to pick any simple topology or even manually design a specific one that would suit a potential implementation. And since the present work relies on photonic implementations of reservoir computing, this is an important point to keep in mind.

For the rest of this work, we will consider reservoirs with ring-like topology, as depicted in Fig. 1.4. The reason for this choice will be given later, in Sect. 1.2, where we will introduce the experimental setup and time-multiplexing. It will then become clear that such architecture corresponds naturally to a delay system. It has been shown in [65] that the performance of such a simple and deterministically constructed reservoir is actually comparable to a regular random echo state network.

A possible interconnection matrix $a_{ij}$ corresponding to a ring like topology is

$$a_{ij} = \alpha \begin{pmatrix} 0 & 1 & 0 & 0 & \cdots & 0 \\ 0 & 0 & 1 & 0 & \cdots & 0 \\ 0 & 0 & 0 & 1 & \cdots & 0 \\ \vdots & \vdots & \vdots & \vdots & \ddots & \vdots \\ 0 & 0 & 0 & 0 & \cdots & 1 \\ 1 & 0 & 0 & 0 & \cdots & 0 \end{pmatrix}, \tag{1.3}$$

where $\alpha$ is a global scale factor. The physical system we will use corresponds to a slightly different set of equations, which can be written as

$$x_0(n + 1) = f\left(\alpha x_{N-1}(n - 1) + \beta M_0 u(n)\right), \tag{1.4a}$$

$$x_i(n + 1) = f\left(\alpha x_{i-1}(n) + \beta M_i u(n)\right). \tag{1.4b}$$

The difference with Eq. 1.3 corresponds to what the node $x_0(n + 1)$ is connected to: in Eq. 1.3 it is connected to $x_{N-1}(n)$ while in Eq. 1.4 it is connected to $x_{N-1}(n - 1)$. Note that the structure of the $a_{ij}$ matrix is reflected by the dependence of $x_i(n + 1)$ on $x_{i-1}(n)$, while the matrix itself was replaced by a simple coefficient $\alpha$. As it defines the strength of the recurrent part of the network, or feedback, we shall from now on call it feedback gain or feedback attenuation, depending on whether it is superior or inferior to 1, respectively. In a similar way, we have replaced the $b_i$ vector by a global scale factor $\beta$ and a vector $M_i$, drawn from a uniform distribution over the interval $[-1, +1]$. The $M_i$ vector is commonly called input mask, or input weights, as it defines the strengths of the input signal $u(n)$ received by each individual neuron $x_i$. The global scale parameter $\beta$ is therefore called input gain.

The nonlinear function $f$ can be virtually any bounded function. Unbounded functions may work as well, such as the softmax and hardmax functions currently used in deep learning. At the moment of writing these lines, I could not find any systematic study of reservoir performance with different nonlinear functions. A common choice is the hyperbolic function $y = \tanh(x)$ [64]. With hardware implementations, however, the choice of $f$ is rather dictated by the choice of a device with a certain nonlinear transfer function. It could be, for instance, a saturation curve of an optical amplifier [66] or a saturable element [67]. As will be explained later in Sect. 1.2, in this work we will be using exclusively a component with a sine transfer function.

With a sine activation function $f(x) = \sin(x)$, Eqs. 1.4 become

$$x_0(n + 1) = \sin\left(\alpha x_{N-1}(n - 1) + \beta M_0 u(n)\right), \tag{1.5a}$$

$$x_i(n + 1) = \sin\left(\alpha x_{i-1}(n) + \beta M_i u(n)\right). \tag{1.5b}$$

These equations describe the behaviour of the system we will discuss later in this work (see Sect. 1.2). It does not get any more complicated than that!

The output of the network is obtained by a simple linear readout layer, that is, by computing a linear combination of the reservoir states $x_i(n)$ and the readout weights $w_i$ as follows

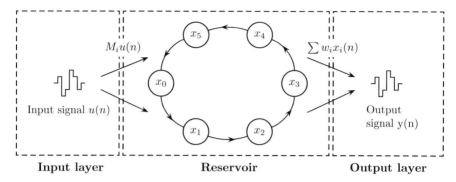

**Fig. 1.4** Schematic representation of a reservoir computer with $N = 6$ nodes. In terms of artificial neural networks, its architecture is composed of a single input neuron (which receives the input signal $u(n)$), one layer of hidden neurons and a single output neuron (which produces the output signal $y(n)$). The configuration of neurons in the hidden layer can be arbitrary, but for ease of hardware implementation we use a ring-like topology. Scheme reprinted from [68]

$$y(n) = \sum_{i=0}^{N-1} w_i x_i(n), \tag{1.6}$$

where $y(n)$ is the temporal output signal. Figure 1.4 gives a graphical overview of the reservoir computer we have just described.

Let us now discuss how the reservoir does what we want it to do. It receives a discrete-time temporal signal $u(n)$ as input, and produces in response another discrete-time temporal signal $y(n)$. With random readout weights $w_i$, this output signal can be anything, and will most likely be something useless. However, the goal is to perform a specific function on the input signal $u(n)$ that would turn it into a desired signal $d(n)$ (examples of such target signals will be given in Sect. 1.1.4). Let us assume that the desired output $d(n)$ is known for several values of input $u(n)$, for instance, $u(1 \ldots 1000)$ and $d(1 \ldots 1000)$ are known. These time series can be used to adjust the readout weights of the system to produce the correct output, i.e. to emulate the specific function we want to execute. Remember the different training approaches we discussed in Sect. 1.1.1? This one falls into the supervised learning category, since we know the inputs and the desired outputs $d(n)$. Training a general RNN would require using the backpropagation algorithm (introduced in Sect. 1.1.2) to tune all the internal connections. Under the reservoir computing paradigm this task is much more simple, as we are only concerned by the readout weights. And since the readout is linear, there is a simple way of training them.

The goal of the training process is to minimise the difference between the actual output of the system $y(n)$ and the desired output $d(n)$ within a certain interval $n \in [1, T]$, where both $u(n)$ and $d(n)$ are known. This interval is commonly referred to as the training interval and its length defines how many "teacher" inputs have been used to optimise the system. The distance measure $D$ between $d(n)$ and $y(n)$ is given by

$$D = \frac{1}{T} \sum_{n=1}^{T} (d(n) - y(n))^2 . \tag{1.7}$$

Since we want to minimise $D$ by tuning the readout weights $w_j$, taking the derivative of $D$ with respect to $w_j$ yields zero

$$0 = \frac{\partial}{\partial w_j} D = \frac{\partial}{\partial w_j} \frac{1}{T} \sum_{n=1}^{T} (d(n) - y(n))^2 . \tag{1.8}$$

Let us develop the right-hand side. The desired output $d(n)$ does not depend on $w_j$, but the output $y(n)$ does, and following Eq. 1.6, we obtain

$$\frac{\partial}{\partial w_j} y(n) = x_j(n). \tag{1.9}$$

Inserting Eq. 1.9 into Eq. 1.8 and expanding the parenthesis gives

$$0 = \frac{\partial}{\partial w_j} \frac{1}{T} \sum_{n=1}^{T} \left(d^2(n) - 2d(n)y(n) + y^2(n)\right), \tag{1.10a}$$

$$= \frac{1}{T} \sum_{n=1}^{T} \left(-2d(n)x_j(n) + 2x_j(n)y(n)\right), \tag{1.10b}$$

$$= \frac{1}{T} \sum_{n=1}^{T} \left(\sum_{i=0}^{N-1} w_i x_i(n)x_j(n) - x_j(n)d(n)\right). \tag{1.10c}$$

And here we obtain a system of linear equations

$$R_{ij} \cdot w_i - P_j = 0 \tag{1.11}$$

for the readout weights $w_i$, where

$$R_{ij} = \frac{1}{T} \sum_{n=1}^{T} x_i(n)x_j(n) \tag{1.12}$$

is the correlation matrix, and

$$P_j = \frac{1}{T} \sum_{n=1}^{T} x_j(n)d(n) \tag{1.13}$$

is the cross-correlation vector. The solution of this system is given by

$$w_i = \sum_{j=0}^{N-1} R_{ij}^{-1} P_j \tag{1.14}$$

and thus, the training of the reservoir computers boils down to the inversion of the correlation matrix $R_{ij}$.

The above problem—minimsation of the distance $D$ (Eq. 1.7) with respect to the unknowns $w$—can be viewed as minimisation of a problem of the form

$$\|Ax - b\|^2, \tag{1.15}$$

where $A$ is a matrix, $x$, $b$ are vectors and $\| \cdot \|$ is the Euclidean norm. Solving Eq. 1.15 is equivalent to finding a solution to a linear system of the kind

$$Ax = b. \tag{1.16}$$

The standard approach to solving such a system is to invert $A$ using the ordinary least squares algorithm [69]. However, in some cases, the problem $Ax = b$ is ill-posed. That is, no $x$ satisfies the equation, or more than one $x$ does, or the solution $x$ has very large values, which makes it unstable with respect to small variations of $A$ or $b$. All these problems arise when the matrix $A$ has small or vanishing eigenvalues. In such cases, using ordinary least squares leads to an overdetermined (over-fitted) or underdetermined (under-fitted) solution. The most common method for regularisation of ill-posed problems is the Tikhonov regularisation [70], also known as ridge regression or weight decay in the machine learning field. The method consists in adding a regularisation term

$$\|Ax - b\|^2 + \|\Gamma x\|^2, \tag{1.17}$$

where $\Gamma$ is a suitably chosen Tikhonov matrix. In many cases, it is chosen as a multiple of the identity matrix $\Gamma = \alpha \mathbb{I}$, with a fixed coefficient $\alpha$. The solution is now $x = (A + \alpha \mathbb{I})^{-1} b$, which is better posed than the original problem. That is, such regularisation gives preference to solutions with smaller norms. It is mostly used in simulations, as experimental noise already does a good job of preventing overfitting in physical implementations. Typically, we set $\alpha \in [10^{-9}, 10^{-1}]$, depending on the reservoir size and the task.

### 1.1.4  Benchmark Tasks

We have just shown that the training of a reservoir computer requires the knowledge of the target signal $d(n)$. In simple words, it is the output we want the system to produce. And this output depends on the task we want the system to perform. This section presents two of the most popular benchmark tasks used to test experimental reservoir computers. Many other tasks can be found in the the literature, but they are out of scope of this thesis, simply because I did not consider them in my experiments. More advanced tasks, such as VARDEL [71] and chaotic time series prediction will be introduced in Chap. 3 and Chap. 4, respectively, as they require important improvements of the Reservoir Computer scheme in order to be solved.

### 1.1.4.1  Wireless Channel Equalisation

This task is based on a real-life situation, depicted in Fig. 1.5. Consider a wireless transmission between an emitter and a receiver. These could be, for instance, a message sent from a broadcast satellite to a ground station, or from a ground station to a personal mobile device. The message arrives to its destination altered by noise and various distortions. The possible causes are (a) interference between different echos of the message, propagating through different paths and thus arriving at the receiver at different moments, (b) imperfect behaviour of the hardware and (c) noise, captured at any stage of the transmission. For this reason, an equaliser is placed at the receiving end to recover the original message. Multiple digital algorithms have

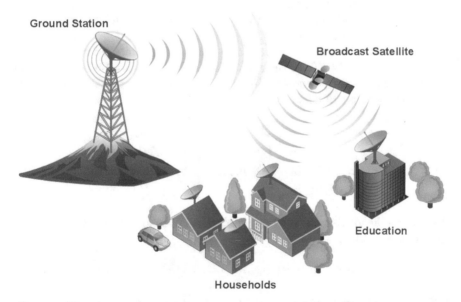

**Fig. 1.5** Wireless channels are omnipresent in our everyday life. The cell phones transmit call and texts to ground stations, that in turn, transfer data to satellites. Smartphones exchange data through Wi-Fi, 3G/4G and Bluetooth. Emergency services, law enforcement agents, taxi drivers communicate with radios. Music and latest news are broadcast to our cars so we do not get bored while sitting in traffic. The list is too long to fit into this page. All these channels are imperfect and the received message often needs to be corrected in order to be readable. This is done by equalisers at the receiving and of the transmission chain. Image reprinted from ConceptDraw

been implemented to perform this task. However, the increasing demand for higher bandwidths requires very fast Analogue-to-Digital Converters (ADCs) that have to follow the high bandwidth of the channel with sufficient resolution to sample correctly the distorted signal [72]. Current manufacturing techniques allow producing fast ADCs with low resolution, or slow ones with high resolution, obtaining both being very costly. This is where analogue equalisers, such as an opto-electronic reservoir computer, become interesting, as they could equalise the signal before the ADC and significantly reduce the required resolution of the converters, thus potentially cutting costs and power consumption [73–75]. Moreover, optical devices may outperform digital devices in terms of processing speed [73, 76]. It can for instance be shown that reservoir computing implementations can reach comparable performance to other digital algorithms (namely, the Volterra filter [77]) for equalisation of a nonlinear satellite communication channel [78].

To emulate a wireless transmission, one starts by generating a message, usually composed of random symbols. The message is then fed through a model of the channel that adds the alterations, caused by transmission, and noise, thus producing a signal that would have been captured at the receiver end. Since the goal of the task is to recover the clean message from the distorted one, the former becomes the target signal $d(n)$ for the reservoir computer, while the latter will be used as the input signal $u(n)$. To avoid confusions, the reader should keep in mind that $d(n)$ is both the target signal for the RC and input signal for the channel model, while $u(n)$ is the output of the channel, but the input for the RC.

The specifics of this task have been defined in [79] and introduced to the reservoir computing field in [62]. The channel input signal $d(n)$ contains 2-bit symbols with values picked randomly from $\{-3, -1, 1, 3\}$. The channel is modelled by a linear system with memory of length 10

$$
\begin{aligned}
q(n) = {} & 0.08d(n+2) - 0.12d(n+1) + d(n) \\
& + 0.18d(n-1) - 0.1d(n-2) + 0.091d(n-3) \\
& - 0.05d(n-4) + 0.04d(n-5) + 0.03d(n-6) \\
& + 0.01d(n-7),
\end{aligned}
\tag{1.18}
$$

that mimics the interference between different echos, followed by an instantaneous memoryless nonlinearity

$$
u(n) = q(n) + 0.036q^2(n) - 0.011q^3(n) + \nu(n),
\tag{1.19}
$$

that replicates the nonlinear behaviour of a signal amplifier at the emitting point, where $u(n)$ is the channel output signal and $\nu(n)$ is the added Gaussian noise. The reservoir computer has to restore the clean signal $d(n)$ from the distorted noisy signal $u(n)$. The performance is measured in terms of wrongly reconstructed symbols, called the Symbol Error Rate (SER).

Note that although the input signal $d(n)$ has a symmetric symbol distribution around 0, the output signal $u(n)$ loses this property, with the symbols lying within the $[-2.8, 4.5]$ interval. The equaliser must take this shift into account and correct the symbol distribution properly.

### 1.1.4.2 NARMA10

This task is the nonlinear version of the Autoregressive-Moving-Average model (ARMA) of order 10, hence NARMA10. The original ARMA model, introduced in [80], consists of two parts: an autoregressive part and a moving average part [81]. The model is suitable for description of systems that combine a series of unobserved shocks (the moving average part) as well as their own behaviour (the autoregressive part). Stock market prices is a good example of such a system.

NARMA10 seems to be a more complex task than channel equalisation. That is, the first opto-electronic reservoir computer, reported by our team [82], achieved very good results on the channel equalisation, that were only slightly improved since then in subsequent experiments. The first experimental results on NARMA10, however, presented in [82] were surpassed in several works, such as e.g. [83] and [84]. The latter will be presented in Chap. 3.

The basic idea of the NARMA10 task is the emulation of a nonlinear system of order 10, hence the name of the task. Other system orders are used, but we will not consider them here. The input signal $u(n)$ is drawn randomly from a uniform distribution over the interval $[0, 0.5]$. The target output $d(n)$ is defined by the following equation

$$d(n + 1) = 0.3d(n) + 0.05d(n) \left( \sum_{i=0}^{9} d(n - i) \right) + 1.5u(n - 9)u(n) + 0.1.$$

(1.20)

Since the reservoir does not produce $d(n)$ exactly, its performance is measured in terms of an error metric. We use the Normalised Mean Square Error (NMSE), given by

$$\text{NMSE} = \frac{\langle (y(n) - d(n))^2 \rangle}{\langle (d(n) - \langle d(n) \rangle)^2 \rangle},$$

(1.21)

where $\langle . \rangle$ is an average over time. A perfect match yields NMSE $= 0$, while a completely off-target output gives a NMSE of 1.

## 1.2 Hardware Implementations: Opto-Electronic Delay Systems

Now that we have covered the key theoretical aspects of reservoir computing, we may address the following question: how does one implement such networks in hardware. This can be done in numerous ways, going from following the idea to the letter, i.e.

with a bucket filled with water [1], up to complex electronic, acoustic and optical solutions [66, 82, 83, 85–93]. The length of the list shows the abundant interest that RC has received from experimental researchers.[3]

In this work, however, we will focus in particular on one such implementation: the first opto-electronic reservoir reported by the OPERA-Photonique group. It combines optics and electronics for a high-speed system. All my experiments were based on this setup, with several add-ons. Therefore, a good explanation of its working principle would not be a luxury. In this section we discuss every component of the setup, how they work all together and how it can be used to process information under the RC paradigm.

## 1.2.1  Time-Multiplexing

As have been explained in Sect. 1.1.3, a reservoir is a network of neurons. Each neuron evolves in time following the activation function. In hardware implementations, this function could be processed by a device, or a dedicated component of the setup: an array of transistors, for instance, or a sequence of operations performed by a microprocessor. Some devices can be made to update multiple neurons in parallel, that is, their operating principle allows for multiple physical or virtual inputs and outputs (e.g. the parallel frequency-multiplexed scheme proposed in [94]). Others, such as the light intensity modulator, used in this thesis (it will be presented in Sect. 1.2.4), can only process one neuron at a time. This means that, in principle, one needs $N$ such modulators, one for each neuron. And since these devices are not cheap, the price of the setup becomes a big problem.

The solution to this issue relies in a careful analysis of Eqs. 1.5. In fact, one may notice that neurons $x_i$ do not need to be updated simultaneously. Since each neuron $x_i$ only depends on one neighbour $x_{i-1}$, they can be updated in a ordered way, that is, one after another. This simple idea allows to replace $N$ activation function components by just one, that would process the queue of neurons $x_0, \ldots, x_N$ at each timestep $n$. Such procedure is commonly called time-multiplexing, as instead of processing all neurons simultaneously, in parallel, they are stacked in a queue, or in other words, time-multiplexed.

Figure 1.6 illustrates the above idea. In order to update the states of all the neurons using only one instance of the activation function, the neurons need to be stacked in a queue. This can be achieved by defining a piecewise constant function of time, with each constant interval corresponding to the value of a certain neuron $x_i(n)$ of the network at time $n$. The output of the function gives the updated states of the neurons at the next timestep $n + 1$, in the form of a new piecewise constant function. To avoid misunderstanding, care should be taken not to confuse the physical time of

---

[3]Although I cannot guarantee the completeness of this list, I did my best to cite all experimental setups known at the moment of writing these lines.

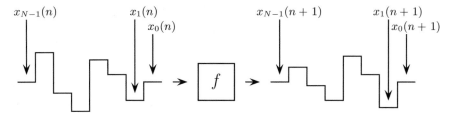

**Fig. 1.6** The basics of time-multiplexing. To update the neural network with one instance of the activation function, the states are encoded into a piecewise constant function of time, where the constant intervals contain the values of the neurons $x_i(n)$ at time $n$. The signal is fed through the activation function block that outputs another piecewise constant signal, containing updated values of neurons $x_i(n+1)$

the piecewise signal, and the discrete time $n$. While the latter is indeed called time for convenience, it is no more than a basic index.

Time-multiplexing thus allows to significantly simplify experimental implementations of reservoir computing. However, this is not the only way to proceed, and the idea itself is far from being flawless. The processing speed of the activation function component defines how fast it can update each neuron. This means that implementing large reservoirs would results in very long queues that would take considerable time to process. In other words, the scalability of the setup becomes an issue. An alternative and very promising way is to encode the reservoir states into different frequencies of a polychromatic light beam. In that case, the network size is only limited by the number of different frequencies that could be created, and this number can go up to hundreds of thousands with particular light sources. And since the frequencies propagate all together, in parallel, all neurons could be processed in parallel. This idea is being studied in our lab at the moment of writing these lines, and I refer the reader to the papers published so far [94, 95].

## 1.2.2 Conceptual Setup

We have just covered how time-multiplexing allows to implement the activation function with just one component instead of $N$. But there is still work to be done. Equations 1.5 also contain a sum of the input signal with previous states of the network. This requires, in principle, a memory block to store the past values. A different approach consists in using a delay system, and in this section we will explain how to do that.

Figure 1.7 illustrates a conceptual setup that implements Eqs. 1.5 with a delay loop. Time-multiplexing is used here not only to process all the states with one instance of the activation function, but also to store previous values in the delay line. The concept of storage may be misguiding at first, as the previous values are not "sitting" in some place, waiting for them to be called, as it would happen in a memory block.

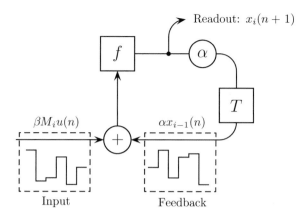

**Fig. 1.7** Conceptual scheme of a hardware setup implementing Eqs. 1.5. The activation function block $f$ outputs the updated reservoir states, which then propagate through the delay loop. The former consists of an amplifier (or attenuator) that applies the feedback gain (or attenuation) $\alpha$ and the delay $T$. The adder sums the input signal from the left with the feedback signal from the right. The inputs and the neuron values are encoded into piecewise constant signals through time-multiplexing. The duration of the delay $T$ is precisely set so that the feedback neurons $x_{i-1}$ are added to the right inputs $M_i u$

Instead, they are constantly on the move, as they propagate through the delay line in the form of a piecewise constant analogue signal. The length of the delay line is chosen precisely so that the feedback signal comes back when it is needed, exactly at the right time (the "right time" will be defined more precisely in the following paragraphs). And that is the beauty of this simple idea [86].

Let us overview, step by step, what happens in this conceptual setup. Suppose that, in the beginning, the system is idle and no signal is present in the delay line. The inputs start arriving from the left as a piecewise constant signal. Each input value $u(0), u(1), \ldots, u(n)$ is multiplied by the input mask $M_i$, which gives $N$ values $u(n)M_i$ for each $n$. In other words, the constant intervals of the input signal correspond to the following values: $u(0)M_0, u(0)M_1, \ldots, u(0)M_{N-1}, u(1)M_0, u(1)M_1, \ldots$. The input signal is first processed by the summation block that adds the input and feedback signals. As we have said above, no feedback signal is present in the loop yet, so the input signal moves upwards on the scheme unaltered. It is then processed by the activation function block that outputs the reservoir states $x_i$ at time $n = 1$. The output of this block is also the most suitable place to read the state of the neural work. That is why the readout arrow on the scheme is located here.

At this point, the updated reservoir states begin their journey in the delay loop on the right-hand side of the scheme. They are first fed through a component that applies the feedback scaling factor, that is, multiplies all the values by the feedback gain $\alpha$. Next comes the delay $T$, accurately chosen so that the feedback signal comes back to the summation block at the right time. Specifically, the neuron $x_{i-1}(n)$ arrives to the sum from the right at the same time as the input $u(n)M_i$, so as to satisfy Eq. 1.5.

Here is a simple example to illustrate the whole process. Consider the first input $u(0)M_0$ that enters the system. Since no feedback is present at the adder yet, it becomes the reservoir state $x_0(1)$ at the output of the activation function. Then, it propagates through the delay loop and arrives at the summation block from the right precisely at the moment when the input $u(1)M_1$ enters the system from the left.

### 1.2.3 Desynchronisation

Before we move to the actual experimental setup, let me say a few words on the principle of desynchronisation, that we used in the above process without actually naming it. From what was explained above, a reservoir state $x_{i-1}(n)$ is summed up with an input value $u(n)M_i$. This may seem counter-intuitive, as one may want to combine it with $u(n)M_{i-1}$ to match the indexes. However, there are a few reasons to mismatch them (and, by the way, this is why this approach is called desynchronisation). First, this is done to satisfy Eqs. 1.5. One may argue that this is not the actual reason, as these equation were derived from the ring-like topology that we created artificially. The genuine reason for desynchronising the system is to create interaction between the neurons.

Imagine that, with synchronous indexes, input $M_i u(n)$ is summed with the feedback $x_i(n)$. That means that the neuron $x_i(n)$, through the course of its evolution from timestep 0 to timestep $n$, has only seen input values $u(n)M_i$ with index $i$, and only its own previous values $x_i(n-1), \ldots, x_i(0)$. Such a system is no longer a network of neurons, but a mere set of independent variables. An important property of a neural network is the ability of the neurons to exchange information between them.

To wrap up, desynchronisation is a way to interconnect the neurons within the network. It is important to note that this is not the only approach. In [87], practically the same delay system is run synchronously. The interconnections are created by an added low-pass filter that links the reservoir states together, as its output depends on current and past input values.

### 1.2.4 Experimental Setup

We can now make the final step towards the experimental setup, schematised in Fig. 1.8. Although this setup is the core part of all of my experiments, it is not the novelty of my work, as it has been designed before I joined the lab [82]. For this reason, I present it here, in the introductory chapter, alongside all other concepts that were well known and established before I started my research.

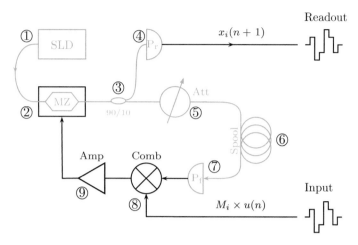

**Fig. 1.8** Schematic representation of the photonic reservoir, introduced in [82]. It contains a light source (SLD), a Mach-Zehnder intensity modulator (MZ), a 90/10 beam splitter, an optical atten-uator (Att), a fibre spool (Spool), two photodiodes ($P_r$ and $P_f$), a resistive combiner (Comb) and an amplifier (Amp). Optical and electronic components are shown in grey and black, respectively

This experiment is often qualified as opto-electronic, electro-optic, or photonic.[4] In Fig. 1.8, electrical cables and components are drawn in black, and grey lines correspond to optical components and fibre. Remember that the reservoir states are encoded into piecewise constant temporal signals. In this setup, these signals are generated in two different mediums—light and electricity. Thus, several components serve to either generate one of the two mediums or convert the signal from one medium to another.

At first sight, the setup in Fig. 1.8 is quite different from the conceptual design depicted in Fig. 1.7. Let us first go through all the components involved here, and then explain how they do the same thing as the conceptual model. The photonic reservoir computer is composed of the following devices.

① An optical experiment starts with a light source: a SLD (superluminiscent diode) producing broadband light at the standard telecommunication wavelength 1550 nm.

② The light intensity is modulated by the Mach-Zehnder intensity modulator (MZ) that shapes it proportional to the input electrical voltage. In other words, it serves to transfer information from an electrical signal into an optical one.

③ Following the light path in optical fibre, next comes a 90/10 splitter. As its name suggests, it splits the light beam in two fractions with the given ratio.

---

[4]Photonics is quite a tricky term. I am yet to find an established and precise definition and, in my experience, various scientists interpret this concept differently. In the present work, for simplicity, I make no distinction between these three terms.

④ A photodetector, or photodiode ($P_r$) produces an electrical signal proportional to the input optical signal. Its function may been as the opposite of the intensity modulator—to transfer information from an optical signal into an electrical one.

⑤ The function of the optical attenuator (Att) is given explicitly by its name—it attenuates the light intensity by a fixed factor, nothing more.

⑥ The fibre spool (Spool) is a big reel of optical fibre. Its purpose is to delay the signal between the optical attenuator (Att) and the following photodiode ($P_f$). As the speed of light in the standard optical fibre is, roughly, $2 \times 10^8$ m/s, one kilometre of fibre creates a delay of about 5 μs. As will be shown below, this order of magnitude is sufficient for this setup.

⑦ A second photodiode ($P_f$), identical to ($P_r$), converts the delayed optical signal into voltage.

⑧ This voltage is added up with an external electrical signal, containing the inputs to the system.

⑨ Finally, the newly produced voltage is amplified by an electrical amplifier (Amp), as the Mach-Zehnder modulator "expects" input voltages much larger than the photodiode $P_f$ can generate.

Table 1.1 lists the exact device models used for this setup with their main characteristics.

The key element of the setup is the Mach-Zehnder intensity modulator, since it carries out the activation function of the neurons. The light intensity at its output is given by [82]

$$I(t) = \frac{I_0}{2} + \frac{I_0}{2} \sin\left(\frac{\pi V(t)}{V_\pi} + \phi\right), \tag{1.22}$$

where $I_0$ is the input light intensity and $V(t)$ is the time-dependent voltage driving the modulator. The bias $\phi$ can be adjusted by applying a DC voltage $V_\phi$ to the modulator. The constant voltage $V_\pi$ is an intrinsic characteristic of the modulator, that corresponds to the voltage needed to go from a maximum to the next minimum of light intensity at the output of the modulator (in our case, $V_\pi \approx 4.5$ V). The transfer function of the modulator is the reason why we use a sine activation function, as have been mentioned previously.

The reservoir states $x_i(n)$ can be both positive and negative. Hence, the voltage $V(t)$, driving the modulator, consists of positive and negative values. However, the Mach-Zehnder outputs a modulated light intensity that only holds positive values. Therefore, the output voltage of the feedback photodiode $P_f$ is strictly positive. It can be broken down into a DC voltage $V_{DC}$, proportional to the light intensity $I_0/2$, and an AC voltage $V_{RF}$, proportional to the intensity fluctuations around the mean value. The DC voltage is cut off by the high-pass filter of the amplifier, so that only $V_{RF}$ is amplified and used to drive the modulator $V(t) \sim V_{RF}$. The filter thus allows both positive and negative reservoir states, despite the fact that they are encoded into strictly positive light intensity. More details on this aspect can be found in the Supplementary Material of [82].

**Table 1.1**  Main components of the opto-electronic reservoir, schematised in Fig. 1.8

| Component | Main characteristics |
|---|---|
| Light source | Thorlabs SLD1550P-A40<br>• centre wavelength: 1550 nm<br>• FWHM: 33 nm<br>• maximum output power: 40 mW |
| Intensity modulator | EOSPACE AX-2X2-0MSS-12<br>• bandwidth: > 10 GHz<br>• $V_\pi$ (at 1 GHz): 4.5 V |
| Photodiodes | TTI TIA-525I<br>• bandwidth: DC to 35 MHz or DC to 125 MHz (switchable)<br>• maximum output voltage: 2 $V_{p\text{-}p}$<br>• maximum linear input: 1.2 mW |
| Optical attenuator | Agilent 81571A<br>• attenuation range: $0 - 60$ dB<br>• resolution: 0.001 dB |
| Fibre spool | Standard SMF-28e fibre<br>• length: approx. 1.5 km |
| Resistive combiner | Home-made star format power splitter<br>• resistors (3x): $16.7\Omega$ |
| Amplifier | Mini Circuits ZHL-32A+<br>• gain: 25 dB<br>• bandwidth: $0.05 - 130$ MHz<br>• maximum input: 2 $V_{p\text{-}p}$ at $50\Omega$ |

We will now discuss the operating principle of the entire setup. We will proceed in the same manner as we did with the conceptual setup, so as to highlight the similarity between the schemes. To start, let us suppose that the experiment is idle (that is, no signals are present at any point) at the moment when the first input comes into the reservoir.

The inputs arrive into the system as an electrical signal (the bottom right corner on the scheme). This signal is the same piecewise constant function containing the input signal $u(n)$, multiplied by the input mask $M_i$. The resistive combiner (Comb) sums the input and feedback signals. Since the latter is null,[5] the input signal alone is amplified (Amp) and applied to the intensity modulator (MZ), that shapes the light intensity into the same piecewise constant function, proportional to the sine of the input electrical signal. In other words, the input signal $M_i u(n)$ is passed through the modulator transfer function (here, $\sin(x)$) and transferred from voltage to light intensity, so that the optical output contains values $\sin(M_i u(n))$. This optical signal is then split in two. 10% are sent to the readout photodiode $P_r$. Similar to the readout arrow in the conceptual setup, the readout photodiode allows to capture the reservoir

---

[5]Technically, it is not null: the SLD is emitting light, hence the DC voltage $V_{DC} \sim I_0/2$ is present. But we can ignore it, since it is filtered by the amplifier.

states, as it produces an electrical signal proportional to $x_i(n+1)$. At this point, $n = 0$ and $x_i(1) = \sin(M_i u(0))$ since there is no feedback signal yet in the reservoir. The 90% of the optical signal make their way into the optical attenuator (Att), where the feedback attenuation ($\alpha$ in Eqs. 1.5) is applied. The resulting feedback signal then propagates through the delay line.[6] Finally, the feedback neurons are transferred back from light intensity into voltage by the feedback photodiode ($P_f$). To illustrate the process in motion, consider the $N + 1$-st input $M_0 u(1)$ passing through the combiner. No feedback is added to this input, as $x_{N-1}(0)$ is null. However, moments later, as the next ($N + 2$-st) input $M_1 u(1)$ enters the combiner, it is being added to the first feedback value $x_0(1) = \sin(M_0 u(0))$, obtained from the first input $M_0 u(0)$ to the reservoir. Note the mismatch of indexes because of the desynchronisation of the reservoir, as was explained in Eq. 1.2.3.

Accurate choice of delay $T$ is key for precise combination of the input with the feedback. This can be done in two ways. Cutting optical fibre at the desired lengths is very unpractical, so instead of adjusting $T$, we tune the duration of the intervals in the piecewise constant signals. This can be easily achieved with signal generation and acquisition devices. In practice, we start by building a reservoir computer with a certain fibre spool, then measure the delay time $T$ by sending in a spike and estimating the time between its echos on a scope. A basic scope with 60 MHz bandwidth allows to measure $T$ with enough precision. From the number of neurons $N$ that we want to fit into the reservoir, we define the duration of one neuron $\theta = T/N$. In other words, $\theta$ is the duration of each constant step of the piecewise signal. For instance, with 1 km of fibre and $T = 5\mu$ s we can fit $N = 50$ neurons by setting $\theta = 100$ ns. This corresponds to a frequency of 10 MHz. The signals can be generated and recorded by arbitrary waveform generators (AWG) and data acquisition cards, respectively. To get rid of the transients, induced by finite bandwidths of physical devices, the acquired signal can be sampled at a higher frequency, e.g. 200 MHz, and then averaged over 20 samples.

To conclude this section, I list several typical characteristics of the experimental setup. These values are presented for readers willing to accurately reproduce our experiment.

- All electronic inputs and outputs are impedance-matched to $50\Omega$.
- The SLD pump current is set to $250 - 350$ mA, so that the optical power at the readout photodiode does not exceed 1 mW (linear response threshold).
- The input gain is usually set between 0.1 and 0.5 (dimensionless values used in simulations). This roughly corresponds to signals ranging from 25 mV$_{\text{p-p}}$ to 125 mV$_{\text{p-p}}$.
- With input voltages of 120 mV$_{\text{p-p}}$ and higher, the output of the amplifier spans the $V_\pi$ interval $[-4.5$ V$, +4.5$ V$]$.
- The feedback attenuations are typically tuned between 4.5 dB and 6 dB—lower values would put the cavity in a regime where it oscillates spontaneously, which

---

[6]Note that the delay $T$ is the total propagation time from the MZ optical output to its electric input, that is, the full loop. In other words, fibre patch cords and electrical cables also add up to the delay, but their contribution is relatively small.

decreases its performance as a reservoir, while higher values would not provide enough feedback to the reservoir;

- The photodiodes (TTI TIA-525I) carry two amplification stages, with selectable coupling (DC or AC). To avoid saturation of the second amplification stage, both photodiodes are set to AC coupling. The cutoff frequency (100 Hz) is sufficiently low to keep the signal distortion minimal. We also select the minimum 1x gain and maximum bandwidth of 135 MHz.
- The fibre spool of approx. 1.5 km yields a (measured) delay of 7.94μ s. To fit 50 neurons into the reservoir, we set the sampling frequency at 128.4635 MHz (with $\theta = 155.7$ ns) and average the reservoir states over 20 samples.

## 1.3   Field-Programmable Gate Arrays

So far we have discussed the theory behind reservoir computing (Sect. 1.1), and how it could be implemented physically in an opto-electronic experiment (Sect. 1.2). What follows deviates completely from those two topics. This section introduces the true novelty of my work—an amazing device that I had a chance to play with for four years—a FPGA chip. Its outstanding properties, such as high computational speed and intrinsic parallelism, were crucial in most of the experiments presented in this thesis. Therefore, it is natural to devote a section to thoroughly introduce the FPGAs. After a short history lesson, we will discuss the internal structure of the specific FPGA chip I was using, together with the software tools required to program and operate it. The contents of the historical introduction is inspired by [96] with additional information taken from [97] and [98]. The rest of the section is mostly original, with some facts from [99].

### 1.3.1   History

#### 1.3.1.1   From Transistors to Integrated Circuits

Our story begins in 1947 at Bell Labs, when John Bardeen, William Shockley and Walter Brittain invented the first transistor...at least according to Bell Labs legal documents. The first patents on field-effect transistors were issued in 1925 in Canada to Julius Edgar Lilienfeld and in 1939 in Germany to Oskar Heil, although there is no direct evidence that these devices were actually built. Anyways, the 1956 Nobel Prize in Physics for the discovery of the transistor effect was awarded to Shockley, Bardeen and Brattain.

Many consider the transistor to be one of the greatest inventions of the 20th century. Its main applications are amplification and switching of electronic signals, and it plays the role of the main building block of the whole modern electronic industry. But how did we get there?

The first junction transistors were bulky, consumed a lot of power and suffered from various performance issues, such as trapping or scattering of carriers. Operating multiple transistors together in an electronic circuit did not seem possible at that time. However, several technological advances changed the situation completely. In 1959, Dawon Kahng and Martin M. Atalla at Bell Labs invented the metal-oxide-semiconductor field-effect transistor (MOSFET). With a design fundamentally different from the previous bipolar junction transistor, the MOSFET was composed of an insulating layer of silicon dioxide on the surface of a semiconductor (crystalline silicone), with a metallic gate electrode on top. The subsequent progress of clean rooms, reducing contamination to unprecedented levels, and the evolution of photolithography contributed to the development of the Si–SiO$_2$ technology, thus making the MOSFET the most widely used type of transistor in integrated circuits. Let us explore the reasons of such a massive success.

There are two main types of MOSFETs: pFETs and nFETs. They either block or open the current flow depending on what value they receive, 0 or 1. The CMOS (complementary metal-oxide semiconductor) technology, patented in 1963 by Frank Wanlass, took the MOS transistors to the next level. Pairing pFETs with nFETs during the fabrication process of the integrated circuit, so that one in each pair is always off, allowed to significantly cut power consumption and heat dissipation, as the current only flows when the transistors are actually switched.

Several transistors can be combined on a protoboard to obtain a logic gate, e.g. a NAND gate, illustrated in Fig. 1.9. An array of several logic gates, composed of junction transistors, would result in a quite bulky setup. The MOSFET and CMOS technologies allowed to significantly scale the things down by printing large amounts of transistors on relatively small areas of silicon. Such printed boards, called Integrated Circuits (IC), first appeared in 1961. As the transistor printing technology improved further, allowing to fit more and more units on smaller areas, the famous Moore's Law was announced in 1965. It predicts the number of transistors in IC to double every year. It is used for goal setting in industry and research. However, it is doomed to break down very soon. In fact, current photolithography process allows to create chips with features of just 14 nm. Even with improved techniques, it is unclear how much further scaling is possible. At 2 nm, transistors would be just 10 atoms wide, and it is unlikely that they would operate reliably at such a small scale.

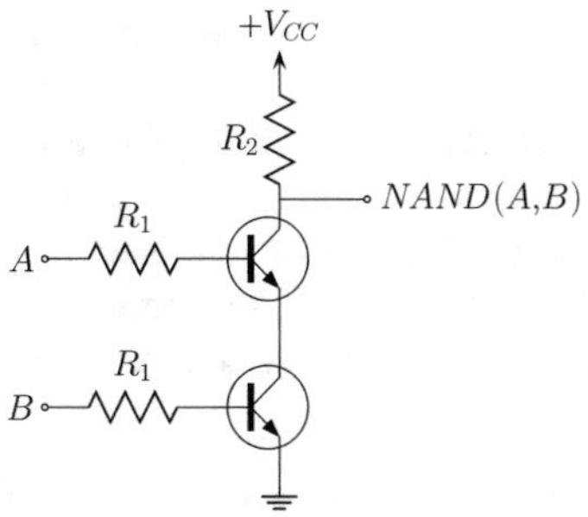

**Fig. 1.9** Logic NAND gate built with two transistors

Moreover, as the transistors are packed ever tighter, dissipating the energy that they use becomes much harder [100, 101].

A simple integrated circuit may consist of several logic gates or a single multiplexer. Modern high-end chips, on the other hand, can contain up to 25 million transistors per square millimetre. The design of such complex ICs is a costly and time-demanding process, with expenses up to multiple tens of millions of dollars. In fact, the circuit has to be conceived to suit all required specifications before being placed on a silicon substrate. This often requires adjusting individual transistors manually, which is virtually impossible for large circuits. Therefore, automatic design software tools play an important role in this design process, that still requires a couple of months of work to get from the design to production. Given the high production costs, this approach is only viable for large orders of chips, such as microprocessors for consumer markets. Smaller projects obviously require a fundamentally different solution to design and test integrated circuits. This is where field-programmable devices—that are alterable by the user himself, not solely in the factory—appear on the stage.

*Programmable logic devices.* A Programmable Logic Device (PLD) is a reconfigurable integrated circuit. Before PLDs were introduced, read-only memory (ROM) was used to perform arbitrary logic functions. However, memory blocks operate much slower than dedicated logic circuits, consume more power and are more expensive in production. In 1970, Texas Instruments developed the first Programmable Logic Array (PLA): the random-access memory (RAM) based device was programmed by altering the metal layer during the production of the IC. In 1978, MMI introduced a Programmable Array Logic (PAL),[7] a device similar to PLA, but instead of two programmable planes, it had one Programmable Read-Only Memory (PROM) array, a fixed OR plane and a programmable AND plane, thus allowing to compute sum of products logic equations with feedback from the outputs. A significant improvement was made by Lattice Semiconductor in 1985 with their Generic Logic Array (GAL), that could be programmed and reprogrammed. It combined CMOS technology with electrically erasable gate technology, making it a high-speed and low-power logic device.

The first logic arrays (PALs and GALs) were only available in small formats of few hundreds of logic gates. Complex Programmable Logic Devices (CPLDs) were introduced for bigger logic circuits by linking several PALs with programmable interconnections. In parallel to the logic arrays, a different type of devices based on gate arrays was being developed, and gave birth to Field-Programmable Gate Arrays (FPGAs). The main difference between a CPLD and a FPGA lies in the architecture. Basically, a logic cell in a CPLD can only be connected to its neighbours, while in a FPGA it can be linked (routed) to any other cell across the chip through programmable interconnections. In other words, a CPLD has programmable logic with stiff connections between cells, while a FPGA offers programmable logic and programmable interconnections. Both CPLDs and FPGAs coexist nowadays—the

---

[7]It seems that the engineers ran out of inspiration when they named their devices! Do not worry if you get lost in all these acronyms, though—we will not use them past this section.

former are limited in size and capabilities and thus used for simpler designs, while the latter are employed for the most complex applications.

How do these devices retain their configuration? A PLD can be seen as a combination of logic and memory units. The former perform individual logic operations and the latter store the connectivity pattern between the cells, given to the device during the configuration process. Different methods have been used through history. The simplest is a silicon antifuse, which works in opposite way to a normal fuse, creating a connection upon destruction, when a voltage is applied. Obviously, such method is indeed field-programmable, but can only be done once. A significant step forward was made with the arrival of electronic memory. At first, Programmable Read-Only Memory (PROM) was used in 1970—it allowed to program the chip fairly easily, but it still was not reprogrammable—once the memory was set, the chip configuration was permanent. Erasable PROM (EPROM) appeared one year later, and brought the valuable advantage of being reprogrammable after erasing its contents with UV-light. Not the most practical solution though, as the EPROM had to be removed from the circuit to be erased. The major improvement was made in 1983 with the invention of EEPROM—Electrically Erasable PROM—that allowed data to be read, erased and rewritten. Still not perfect yet, as it could only be reprogrammed a limited number of times. For this reason, modern FPGAs store their configuration in volatile Static RAM (SRAM), which can be rewritten an unlimited number of times.

To sum up, the need for a simple and inexpensive solution for fast development of integrated electronic circuits—more efficient than placing individual components on a protoboard, and less expensive than following the entire process of IC manufacturing—brought to life several families of programmable logic devices, among which the FPGA—a large array of logic gates (up to two millions), combined with SRAM blocks to store the internal connections.

## 1.3.2   Market and Applications

Nowadays, FPGA market is dominated by two players: Xilinx and Altera. Together, they hold 90% of the market share. Other companies, such as Lattice Semiconductor, Actel, Achronix and Tabula provide more specialised chips with unique features.

Xilinx offers several FPGA families with a wide range of applications. The Artix family holds the entry level chips with lowest cost and power consumption and small form-factor packaging.[8] The Spartan family targets cost-sensitive and high-volume requests. The Kintex family is optimised for the best price-performance. Virtex is the top-level family, with the highest system performance and capacity. Several generations of devices exist within each family. The most recent devices, at the moment of writing those lines, are Family 7 chips. In this work I used a Family 6 device, since it was the latest device on the market in 2013, when it was purchased.

---

[8] An Artix evaluation board can be purchased for as low as $100.

Altera also offers several FPGA families, based on application needs. Stra-tix devices offer the highest performance and density, Arria is a family of mid-range FPGAs, and Cyclone series are the company's lowest cost, lowest power chips.

At first, FPGAs were mainly used to connect electronic components together, such as bus controllers or processors. As they became larger and faster, with lower cost per logic gate, their application landscape changed dramatically. Modern FPGAs are capable of replacing ASIC chips and are more and more frequently found in consumer electronic devices. Gone are the days when they were too expensive for high-volume productions. Furthermore, FPGA chips became reliable enough for critical applications in space, military and medical fields. They are now commonly used for high-performance signal processing, replacing multiple dedicated DSP processors. The latest trend is the System on Chip (SoC) platforms, consisting of multi-core processors and high-capacity logic devices, all on a single chip. The idea is to benefit from the high computational power of the FPGA without the complex design process. The user would only interacts with the processor, while the latter could call the FPGA for support, for instance, to increase its performance on specific tasks by delegating highly repetitive routines to the FPGA.

### *1.3.3   Xilinx Virtex 6: Architecture and Operation*

With the historical background set, let us take a closer look inside a FPGA chip. This section is aimed at users with high interest in FPGA technology. If you are not one of them, you can skip this part—you will miss a lot of interesting facts, but this will not prevent you from understanding the experiments described in the next chapters.

At this point, describing a "general" chip makes no sense, as each manufacturer, be it Altera or Xilinx, designs its chips in different ways with different architectures. We will thus focus on the FPGA chip I have been using throughout my thesis: the Xilinx Virtex 6. Some parts of this section remain valid for all FPGAs, but others are unique features only present on Xilinx devices. The contents of this section was inspired from [99].

Up to now, we simplified the internal structure of a FPGA to (a) a large number of logic gates and (b) blocks of memory to store the interconnections. Modern FPGAs are much more complex than that and contain numerous additional components. This is what we are going to overview here.

*Logic blocks*. The primary function of a FPGA is to perform logic operations, such as AND, OR, NOT, NAND, NOR, XOR, XNOR and more complex combinations of these. Instead of replicating thousands of individual logic gates, FPGA manufacturers make use of a more advanced logic block, called Look-Up Table, or LUT. A general LUT, with $n$ inputs and one output, can encode any $n$-input Boolean function by modelling it as a truth table. That way, several individual logic gates can be replaced by a single logic block. A Virtex 6 LUT has 6 inputs. The output of a LUT may be stored in a register (a simple circuit acting as a one-bit memory) in order to implement sequential logic. Four such LUTs are grouped in a Slice, that also contains eight registers, multiplexers (used to select inputs) and arithmetic carry logic (used to perform arithmetical operations). Two slices form a Configurable Logic Block, or CLB—a feature-rich circuit capable of implementing various logic functions. A FPGA chip from Virtex-6 family may contain from $10^4$ up to $10^5$ CLBs [102]. Obviously, each CLB contains more features than may be required. During the implementation process, mapping tools select which components of each CLB are used, and the others are bypassed or left unconnected from the circuit.

*Routing resources.* The thousands of CLBs (and other components, described below) of a FPGA can be connected in various ways by the user. Therefore, a sufficient number of routing channels (or paths) should be present on the chip in order to make any connectivity pattern possible. Surprisingly, these routing resources occupy most of the physical space on the silicon plate! In fact, the logic blocks can be pictured as small islands floating in a vast sea of routing resources. The channels are arranged in vertical and horizontal grids. As they are printed very densely on the chip, they may resemble a fabric. That is why the combination of FPGA logic and routing resources is frequently called FPGA fabric.

*Clocking resources.* To ensure consistent and predictable outcomes, most sequential logic is synchronous (or clocked), thus requiring a clock signal driving the registers. Reliable clock generation and distribution across the distant corners of the chip is a challenge of extreme importance. No matter how efficient the logic is, if different elements are not synchronised properly, the circuit is doomed to fail. This potential issue is addressed by a dedicated network, used solely for clock distribution. The complex network consists of five parallel lanes used for different purposes, such as local or global clock distribution. These lanes are designed to drive multiple logic blocks at once, with the shortest propagation delay, thus limiting desynchronisation between distant components.

A FPGA chip does not have a built-in clock generator, it has to be clocked from a external oscillator. On the other hand, it accommodates specific circuitry capable of altering the phase and frequency of an incoming clock. Such blocks, called Mixed-Mode Clock Managers (MMCMs), can be used as clock dividers or multipliers.

*Memory.* Virtex-6 FPGAs typically have several hundreds of small blocks of RAM spread across the chip, each containing up to 36 kilobits of memory. An interesting feature of Xilinx devices is another type of memory, called distributed RAM. The data can be stored in the LUTs of CLBs, each LUT being capable of holding 64 bits. Between 25% and 50% of all slices can be used as distributed memory. This allows to keep small amounts of data for direct access.

*DSP Slices.* FPGA applications to digital signal processing (DSP) may require fast and efficient execution of a series of arithmetical operations, such as multiplication. Addition and subtraction operations can be implemented fairly easily with bit-logic, but this is far more complex with multiplication. Basically, a multiplication is nothing more than a sequence of additions, but the main concern is that it grows quickly with the sizes of multiplicands. For instance, multiplying two 16-bit numbers requires 16 16-bit adders. To perform these operations more efficiently, Virtex-6 devices are built with several hundreds (768 in our particular chip) dedicated DSP slices. Each one of them contains a full 25-bit by 18-bit multiplier, with several additional features,

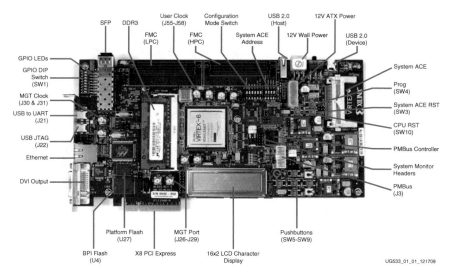

**Fig. 1.10** Xilinx ML605 evaluation board containing the Virtex-6 FPGA chip and enough accessories to employ most of its features. Reprinted from the board getting started guide [104]

such as pre-adders and accumulators [103]. A DSP slice does not compute faster than regular logic. However, it allows to perform multiplication with a single component, saving logic and routing resources for other purposes.

*Xilinx ML605 evaluation board.* For convenience, end-users are offered the option to purchase a FPGA chip that is mounted on a motherboard-like electronic circuit, called evaluation board. That way, the chip comes powered, clocked and ready to use. Moreover, the evaluation board adds useful features and expands the connectivity of the FPGA. Our Virtex 6 came along with a ML605 evaluation board, depicted in Fig. 1.10. It includes switches, LEDs, pushbuttons, a small LCD display, a DDR3 RAM bar, as well as many connection ports, such as PCI Express, DVI, USB, UART and FMC. Some of these features were extensively used in our experiments and will be discussed later.

### 1.3.4  Design Flow and Implementation Tools

The creation of a working hardware implementation is a lengthy process. In this section I will outline the most important steps.

**Coding**:   First, one has to write down (or code) the idea in a specific hardware description language (HDL). There are two choices here: VHDL—Very High Speed Integrated Circuit (VHSIC) Hardware Description Language, strong typed, Ada-like language with annoyingly long syntax, or Verilog—weakly typed, C-like language with more user-friendly syntax. Writing in those languages is

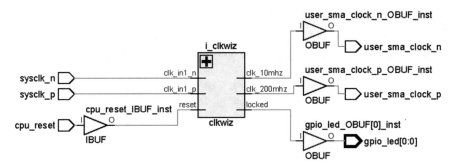

**Fig. 1.11** Example of schematic produced at the synthesis stage. The VHDL design has been converted into FPGA primitives, such as input and output buffers, shown here. The basic design illustrated here creates two clocks (200 MHz and 10 MHz) from the input system clock

fundamentally different from coding a regular program intended for a micro-processor: instead of listing instructions in a particular order, the user describes a hardware (hence hardware description language). In fact, the code is never called a "program", since it is not meant to be executed, but a "design", as it literally describes the design of an electronic circuit. Most HDL instructions are of the kind: "connect output port A of component X to input port B of component Y". Modern HDL compilers are capable of converting more complex structures into logic, such as arithmetical operations or loops. Extreme caution should be taken here, as an HDL loop is not a loop of instructions, and will not be executed as in e.g. C code. Programming in HDL requires a completely different state of mind—"think hardware" advice appears in most books and manuals.

**Simulation**:   Writing the code usually goes in parallel with testing its correct func-tioning. The implementation of a HDL code—that is, the complete process of converting the human-readable code into bits loaded in the FPGA—may take a long time, up to an entire day for very complex designs.[9] Hence, simulating a design is a better option to check the logic in a reasonable time.[10] This can be done with various FPGA emulating programs. In this work, I used iSim, provided by Xilinx with the evaluation board. It allows to generate input and clock signals, and visualise the design behaviour as a time trace of all internal signals. This is particularly handy to verify that all arithmetic and logic operations are executed correctly.

When the simulation results are satisfactory and the design behaves as expected, it is passed to a series of software tools that turn the HDL code into a bitstream that will be loaded into the FPGA. Xilinx Family 6 devices come with ISE Design Suite, a software package containing all the compilers required for implementation. The stages of the process are the following:

---

[9]Implementation times of my designs never exceeded an hour, though.

[10]From a few seconds, up to a minute, in my experience.

**Synthesis**:    The first stage converts the HDL code into a theoretical circuit, using standard components (called primitives) available on the target FPGA chip. The result can be visualised as a Resistor-Transistor Logic (RTL) diagram, illustrated in Fig. 1.11, but cannot be implemented on a physical chip yet.

**Translate**:    The shortest process of all, it merges the synthesised design with the input and output ports, as well as timing and placement constraints. In short, it gathers all design files together before the actual implementation takes place.

**Map**:    The core process that last the longest. It takes into account the specificities of the chip and maps the theoretical design to physical logic blocks of the FPGA. Several algorithms are run to simplify and optimise the mapping. The total FPGA resources utilisation is displayed at the end of the process, that tells how many e.g. CLBs or memory blocks have been used to implement the design.

**Place and Route**:    Although called Place and Route (PAR), this process does not place anything, since it has already been done during the Map process. It only routes the design. In other words, it takes the list of components generated by Map and connects them together, using FPGA routing resources. This is an easy task for small designs, but becomes quite challenging, and takes a while, for large designs. Very large designs may fail at this stage if there is not enough routing paths on the chip to connect the components in the desired manner. After routing, PAR checks the timing closure.

**Timing closure**:    Electrical signals travel at finite speed within the FPGA. Despite the propagation delays being very short, a couple nanoseconds at most, they have to be taken into account for a synchronous design to function properly. And the faster the clock is, the more critical these delays become. Simulation ignores these delays, they are only considered during the hardware implementation stage. Maximum delays are specified by the user, and commonly called "timing constraints". In practice, a specific file lists the clocks used to drive the chip, together with their frequencies. This information allows PAR to check whether the design meets the so-called "timing closure". If it does, and the logic has been checked thoroughly in simulations, then the circuit should work as expected. If it does not, and this happens quite often—the real struggle begins. There is no proven method for meeting time closure. Expert designers can resolve small delay issues using in-depth knowledge of hardware characteristics and implementation software. More complex problems can only be solved by using a slower clock or re-designing the circuit.

**Generate bitstream**:    At last, the design is synthesised, placed, routed, and the timing closure is met. The last stage is the generation of a bitstream, the data that will be loaded into FPGA SRAMs and set its configuration.

A Virtex chip can be configured in various ways. The bitstream can be loaded, for instance, from an on-board non-volatile memory. This consists in loading the bitstream into a specific PROM attached to the evaluation board and set up the FPGA to load its configuration from this memory chip. It can also be programmed by an external microprocessor. In this case, the bitstream is loaded from a e.g. personal computer through the Joint Test Action Group (JTAG) interface—a standard first introduced for debugging of electronic circuits, that is now widely used for loading the configuration file into the device memory. The configuration of the Virtex-6 chip takes about 20 seconds. Once completed, the FPGA is ready to go.

---

This concludes the introductory chapter of my thesis. We have covered most theoretical, experimental, and FPGA-related topics required for understanding the experiments described in the following chapters. A few remaining points, related to particular experiments, will be discussed en route.

# References

1. Fernando, Chrisantha and Sampsa Sojakka. 2003. Pattern recognition in a bucket. In *European conference on artificial life*, 588–597. Springer
2. van Leeuwen, Jan. 1990. *Handbook of theoretical computer science: Algorithms and complexity*. Elsevier
3. Ralston, Anthony, Edwin D. Reilly, and David Hemmendinger. 2000. *Encyclopedia of computer science*. Nature Publishing Group.
4. Reilly, Edwin D. 2003. *Milestones in computer science and information technology*. Greenwood Publishing Group
5. Tucker, Allen B. 2004. *Computer science handbook*. CRC Press
6. Peter, J. 2005. Denning. Is computer science science?". *Communications of the ACM* 48 (4): 27–31.
7. Winston, Patrick Herny. 1984. *Artificial intelligence*. Addison-Wesley
8. Michalski, Ryszard S., Jaime G. Carbonell, and Tom M. Mitchell. 1984. *Machine learning an artificial intelligence approach*. Morgan Kaufmann Publication Incorporated
9. Mitchell, Tim Michael. 1997. *Machine learning*. McGraw-Hill Education
10. Russell, Stuart Jonathan, Peter Norvig, John F Canny, Jitendra M Malik, and Douglas D Edwards. 2003. Artificial intelligenc e: A modern approach. Prentice hall Upper Saddle River.
11. Bishop, Christopher M. 2006. *Pattern recognition and machine learning*. Springer
12. Hastie, Trevor, Robert Tibshirani, and Jerome Friedman. 2013. *The elements of statistical learning: data mining, inference, and prediction*. New York: Springer. 34 Chapter I. Introduction
13. Navada, A., A.N. Ansari, S. Patil, and B.A. Sonkamble. 2011. Overview of use of decision tree algorithms in machine learning. In *2011 IEEE control and system graduate research colloquium*, 37–42. June 2011
14. Kotsiantis, S.B. 2013. Decision trees: a recent overview. *Artificial Intelligence Review* 39 (4): 261–283.
15. Charniak, Eugene. 1991. Bayesian networks without tears. *AI Magazine* 12 (4): 50.
16. Nielsen, Thomas Dyhre, and Finn Verner Jensen. 2009. *Bayesian networks and decision graphs*. Springer Science & Business Media
17. Dasarathy, Belur V. 1991. Nearest neighbor (NN) norms: NN pattern classification techniques
18. Naomi, S. 1992. Altman. An introduction to kernel and nearest-neighbor nonparametric regression. *The American Statistician* 46 (3): 175–185.
19. Shakhnarovich, Gregory, Trevor Darrell, and Piotr Indyk. *Nearest-neighbor methods in learning and vision: Theory and practice (neural information processing)*. The MIT Press
20. Cristianini, Nello, and John Shawe-Taylor. 2000. *An introduction to support vector machines and other kernel-based learning methods*. Cambridge university Press

21. Kecman, Vojislav. 2001. *Learning and soft computing: Support vector ma- chines, neural networks, and fuzzy logic models*. MIT Press
22. Steinwart, Ingo, and Andreas Christmann. 2008. *Support vector machines*. Springer Science & Business Media
23. Salcedo-Sanz, Sancho. 2014. José Luis Rojo-Álvarez, Manel Martínez-Ramón, and Gustavo Camps-Valls. Support vector machines in engineering: An overview. In Wiley Interdisciplinary Reviews. *Data Mining and Knowledge Discovery* 4 (3): 234–267.
24. Hertz, John, Anders Krogh, and Richard G. Palmer. 1991. *Introduction to the theory of neural computation*. Addison-Wesley/Addison Wesley Longman
25. Bishop, Christopher M. 1995. *Neural networks for pattern recognition*. Oxford University Press
26. Gurney, Kevin. 1997. *An introduction to neural networks*. CRC Press
27. Haykin, Symon. 1999. *Neural networks: A comprehensive foundation*
28. Yoshua, Bengio, Aaron Courville, and Pascal Vincent. 2013. Representation learning: A review and new perspectives. *IEEE Transactions on Pattern Analysis and Machine Intelligence* 35 (8): 1798–1828.
29. Deng, Li, and Y. Dong. 2014. Foundations and trends®in signal processing. *Signal Processing* 7: 3–4.
30. LeCun, Yann, Yoshua Bengio, and Geoffrey Hinton. 2015. Deep learning. *Nature* 521 (7553): 436–444.
31. Schmidhuber, Jürgen. 2015. Deep learning in neural networks: An overview. *Neural Networks* 61: 85–117.
32. Hastie, Trevor, Jerome Friedman, and Robert Tibshirani. 2001. Overview of supervised learning. In *The elements of statistical learning*, 9–40. Springer
33. Kotsiantis, Sotiris B., I. Zaharakis, and P. Pintelas. 2007. Supervised machine learning: A review of classification techniques. *In Emerging artificial intelligence applications in computer engineering* 160: 3–24.
34. Sutton, Richard S, Andrew G Barto. 1998. *Reinforcement learning: An introduction*. Vol. 1. 1. MIT Press Cambridge
35. Szepesvári, Csaba. 2009. *Algorithms for reinforcement learning*. Morgan and Claypool
36. Friedman, Jerome, Trevor Hastie, and Robert Tibshirani. 2001. *The elements of statistical learning*. Vol. 1. Springer Series in Sstatistics New York
37. Xu, Lei. 2001. An overview on unsupervised learning from data mining perspective. *Advances in self-organising maps*, 181–209. London: Springer, London.
38. Ghahramani, Zoubin. 2004. Unsupervised learning. In *Advanced lectureson machine learning*. Springer, 72–112.
39. Chapelle, O., B. Schölkopf, and A. Zien. 2006. *Semi-supervised Learning*. Adaptive computation and machine learning: MIT Press.
40. McCulloch, Warren S., and Walter Pitts. 1943. A logical calculus of the ideas immanent in nervous activity. *The bulletin of mathematical biophysics* 5 (4): 115–133.
41. Minsky, Marvin, and Seymour Papert. 1969. *Perceptrons: AnIntroduction to computational geometry*. Cambridge, Mass: MIT Press.
42. Werbos, Paul. 1974. *Beyond regression: New tools for prediction and analysis in the behavioral sciences*
43. Paul, J. 1990. Werbos. Backpropagation through time: What it does and howto do it". *Proceedings of the IEEE* 78 (10): 1550–1560.
44. Alan, L. 1952. Hodgkin and Andrew F Huxley. A quantitative description of membrane current and its application to conduction and excitation in nerve. *The Journal of physiology* 117 (4): 500.
45. FitzHugh, Richard. 1955. Mathematical models of threshold phenomena inthe nerve membrane. *The Bulletin of Mathematical Biophysics* 17: 257–278.
46. Gerstner, Wulfram. 2001. A framework for spiking neuron models: The spikeresponse model. *Handbook of Biological Physics* 4: 469–516.

47. Gerstner, Wulfram, and Kistler, Werner M. 2002. *Spiking neuron models: Single neurons, populations, plasticity.* Cambridge University Press
48. Izhikevich, Eugene M. 2004. Which model to use for cortical spiking neurons? *IEEE transactions on neural networks* 15 (5): 1063–1070.
49. Haykin, Simon. 1998. *Neural networks: A comprehensive foundation.* Prentice Hall. 36 Chapter I. Introduction
50. Rosenblatt, Frank. 1961. *Principles of neurodynamics.* Cornell Aeronautical Lab Inc Buffalo NY: Perceptrons and the theory of brain mechanisms. Tech. rep.
51. Mandic, Danilo P., and Jonathon A. Chambers et al. 2001. *Recurrent neural networks for prediction: Learning algorithms, architectures and stability.* Wiley Online Library
52. Lipton, Z.C., J. Berkowitz, and C. Elkan. 2015. A critical review of recurrent neural networks for sequence learning. In: ArXiv e-prints arXiv:1506.00019 (2015).
53. Turchetti, Claudio. 2004. *Stochastic models of neural networks.* Vol. 102. IOS Press
54. Wong, Eugene. 1991. Stochastic neural networks. *Algorithmica* 6 (1–6): 466.
55. Maass, Wolfgang. 1997. Networks of spiking neurons: The third generation of neural network models. *Neural Networks* 10 (9): 1659–1671.
56. Maass, Wolfgang, and Christopher M Bishop. 2001. *Pulsed neural networks.* MIT Press
57. Ponulak, Filip, and Andrzej Kasiński. 2011. Introduction to spiking neuralnetworks: Information processing, learning and applications. 71: 409–33.
58. Grüning, André and Sander M Bohte. 2014. Principles and challenges: Spiking neural networks. In ESANN.
59. Orr, Mark J.L. etal. 1996. *Introduction to radial basis function networks*
60. Bors, Adrian G. 2001. Introduction of the radial basis function (rbf) networks. *Online symposium for electronics engineers.* 1 (1): 1–7.
61. Wu, Yue, Hui Wang, Biaobiao Zhang, and K-L Du. Using radial basis function networks for function approximation and classification. In *ISRN Applied Mathematics* 2012 (2012).
62. Jaeger, Herbert, and Harald Haas. 2004. Harnessing nonlinearity: Predicting chaotic systems and saving energy in wireless communication. *Science* 304: 78–80.
63. Maass, Wolfgang, Thomas Natschläger, and Henry Markram. 2002. Realtime computing without stable states: A new framework for neural computation based on perturbations. *Neural Computation* 14: 2531–2560.
64. Jaeger, Herbert. 2001. The echo state approach to analysing and training recurrent neural networks—with an Erratum note. In *GMD report* 148
65. Rodan, Ali, and Peter Tino. 2011. Minimum complexity echo state network. *IEEE Transactions on Neural Networks* 22: 131–144.
66. Duport, François, Bendix Schneider, Anteo Smerieri, Marc Haelterman, and Serge Massar. 2012. All-optical reservoir computing. In *Optics Express* 20: 22783–22795. I.4. References 37
67. Dejonckheere, Antoine, François Duport, Anteo Smerieri, Li Fang, Jean-Louis Oudar, Marc Haelterman, and Serge Massar. 2014. All-optical reservoir computer based on saturation of absorption. *Optics Express* 22: 10868–10881.
68. Antonik, Piotr, Marc Haelterman, and Serge Massar. 2017. Brain-inspired photonic signal processor for generating periodic patterns and emulating chaotic systems. In *Physical Review Applied* 7: 054014.
69. Amemiya, Takeshi. 1985. *Advanced econometrics.* Harvard University Press
70. Tikhonov, Andrei Nikolaevich, A.V. Goncharsky, V.V. Stepanov, and Anatoly G. Yagola. 1995. *Numerical methods for the solution of ill-posed problems*, vol. 328. Netherlands: Springer.
71. Hermans, Michiel. 2012. Expanding the theoretical framework of reservoir computing. PhD thesis. Ghent University
72. Singh, Jaspreet, Sandeep Ponnuru, and Upamanyu Madhow. 2009. Multigigabit communication: The ADC bottleneck. In *IEEE international conference on Ultra-Wideband, 2009. ICUWB*, 22–27.IEEE
73. Sobel, David Amory, and Robert W. Brodersen. 2009. A 1 Gb/s mixed-signal baseband analog front-end for a 60 GHz wireless receiver. *IEEE Journal of Solid-State Circuits* 44 (4): 1281–1289.

74. Feng, Xiaodong, Guanghui He, and Jun Ma. 2010. A new approach to reduce the resolution requirement of the ADC for high data rate wireless receivers. In *2010 IEEE 10th international conference on signal processing (ICSP)*, 1565–1568. IEEE
75. Yong, Su-Khiong, Pengfei Xia, and Alberto Valdes-Garcia. 2011. *60 GHz technology for Gbps WLAN and WPAN: from theory to practice*. Wiley
76. Hassan, Khursheed, Theodore S Rappaport, and Jeffrey G Andrews. 2010. Analog equalization for low power 60 GHz receivers in realistic multipath channels. In *2010 IEEE global telecommunications conference (GLOBE-COM 2010)*, 1–5. IEEE.
77. Malone, Jerry, and Mark A. Wickert. 2011. Practical volterra equalizers for wideband satellite communications with twta nonlinearities. In *Digital signal processing workshop and IEEE signal processing education workshop (DSP/SPE), 2011 IEEE*, 48–53. IEEE
78. Bauduin, Marc, Anteo Smerieri, Serge Massar, and François Horlin. 2015. Equalization of the non-linear satellite communication channel with an echo state network. In *Vehicular technology conference (VTC Spring), IEEE 81st*, 1–5. IEEE
79. Mathews, V John, and Junghsi Lee. 1994. Adaptive algorithms for bilinear filtering. In SPIE's 1994 international symposium on optics, imaging, and instrumentation. *International Society for Optics and Photonics*. 317–327.
80. Whitle, Peter. 1951. *Hypothesis testing in time series analysis*. Vol. 4. Almqvist & Wiksells. 38 Chapter I. Introduction
81. Hannan, Edward James. 2009. *Multiple time series*. Vol. 38. Wiley & Sons
82. Paquot, Yvan, François Duport, Anteo Smerieri, Joni Dambre, Benjaminschrauwen, Marc Haelterman, and Serge Massar. 2012. *Optoelectronic reservoir computing. Scientific Reports* 2: 287.
83. Vinckier, Quentin, François Duport, Anteo Smerieri, Kristof Vandoorne, Peter Bienstman, Marc Haelterman, and Serge Massar. 2015. High-performance photonic reservoir computer based on a coherently driven passive cavity. *Optica* 2 (5): 438–446.
84. Hermans, Michiel, Piotr Antonik, Marc Haelterman, and Serge Massar. 2016. Embodiment of learning in electro-optical signal processors. *Physical Review Letters* 117: 128301.
85. Schürmann, Felix, Karlheinz Meier, and Johannes Schemmel. 2004. Edge of chaos computation in mixed-mode VLSI-A Hard liquid. In *NIPS*, 1201–1208.
86. Appeltant, Lennert, Miguel Cornelles Soriano, Guy Van der Sande, JanDanckaert, Serge Massar, Joni Dambre, Benjamin Schrauwen, Claudio R Mirasso, and Ingo Fischer. 2011. Information processing using a single dynamical node as complex system. *Nature Communications* 2: 468.
87. Larger, Laurent, M.C. Soriano, L. Daniel Brunner, Jose M. Appeltant, Luis Pesquera Gutiérrez, Claudio R. Mirasso, and Ingo Fischer. 2012. Photonic information processing beyond Turing: An optoelectronic implementation of reservoir computing. *Optic Express* 20: 3241–3249.
88. Martinenghi, Romain, Sergei Rybalko, Maxime Jacquot, Yanne KouomouChembo, and Laurent Larger. 2012. Photonic nonlinear transient computing with multiple-delay wavelength dynamics. *Physical Review Letters* 108: 244101.
89. Brunner, Daniel, Miguel C. Soriano, Claudio R. Mirasso, and Ingo Fischer. 2013. Parallel photonic information processing at gigabyte per second data rates using transient states. *Nature Communications* 4: 1364.
90. Vandoorne, Kristof, Pauline Mechet, Thomas Van Vaerenbergh, Martin Fiers, Geert Morthier, David Verstraeten, Benjamin Schrauwen, Joni Dambre, and Peter Bienstman. 2014. Experimental demonstration of reservoir computing on a silicon photonics chip. *Nature Communications* 5: 3541.
91. Haynes, Nicholas D., Miguel C. Soriano, David P. Rosin, Ingo Fischer, and Daniel J. Gauthier. 2015. Reservoir computing with a single timedelay autonomous Boolean node. *Physical Review E* 91 (2): 020801.
92. Torrejon, Jacob, Mathieu Riou, Flavio Abreu Araujo, Sumito Tsunegi,Guru Khalsa, Damien Querlioz, Paolo Bortolotti, Vincent Cros, Akio Fukushima, Hitoshi Kubota, et al. 2017. Neuromorphic computing with I.4. References 39nanoscale spintronic oscillators. In arXiv preprint arXiv:1701.07715

93. Larger, Laurent, Antonio Baylón-Fuentes, Romain Martinenghi, Vladimir S. Udaltsov, Yanne K. Chembo, and Maxime Jacquot. 2017. High-speed photonic reservoir computing using a time-delay-based architecture: Million words per second classification. *Physical Review X 7*, 011015.

94. Akrout, Akram, Arno Bouwens, François Duport, Quentin Vinckier, Marc Haelterman, and Serge Massar. 2016. Parallel photonic reservoir computing using frequency multiplexing of neurons. In arXiv:1612.08606

95. Akrout, Akram, Piotr Antonik, Marc Haelterman, and Serge Massar. 2017. Towards autonomous photonic reservoir computer based on frequency parallelism of neurons. *Proceedins SPIE* 10089. 100890S- 100890S-7.

96. Kadric, Edin. 2011. An FPGA implementation for a high-speed optical link with a PCIe interface. PhD thesis

97. Franz, Kaitlyn. 2015. *History of the FPGA*. http://blog.digilentinc.com/history-of-the-fpga/.

98. Wikipedia. Transistor. 2017. http://en.wikipedia.org/wiki/Transistor.

99. Stavinov, Evgeni. 2011. *100 Power tips for FPGA designers*. CreateSpace Independent Publishing Platform

100. Waldrop, M. Mitchell. 2016. The chips are down for Moore's law. *Nature* 530: 144–147.

101. Bright, Peter. 2016. Moore's law really is dead this time. https://arstechnica.com/information-technology/2016/02/moores-law-really-is-dead-this-time/.

102. Virtex-6 Family Overview. 2012. DS150 (v2.4). Xilinx Inc.

103. Virtex-6 FPGA DSP48E1 Slice. 2011. UG369. Xilinx Inc.

104. Getting Started with the Xilinx Virtex-6 FPGA ML605 Evaluation Kit. 2011. UG533 (v1.5). Xilinx Inc..

# Chapter 2
# Online Training of a Photonic Reservoir Computer

This chapter presents the first experiment I performed during my Ph.D.[1] It took me almost a year and half to get going, and six more months to gain full control of the experiment and obtain publishable results. The reason for such a slow start is, without any doubt, the immense complexity of FPGA programming. But let us not focus on the difficulties (we will get to them later, in Sect. 2.7) but rather on the achievements. And as you will find out by the end of this chapter, this first experiment produced some very interesting results, that inspired and shaped my following research projects. But first, let me outline what motivated this research, and what we expected to achieve in the first place.

## 2.1 Introduction

The performance of a reservoir computer greatly relies on the training technique used to compute the readout weights. Offline learning methods, introduced in Sect. 1.1.3 and used up to now in experimental implementations [2–10], provide very good results. However, they start to cause problems in real-time applications, as they require large amounts of data to be transferred from the experiment to the post-processing computer. This operation may take longer than the time it takes the reservoir to process the input sequence [4, 7, 9]. Moreover, offline training only works on time-independent tasks, which is not always the case in real-life applications. The

---

[1]The contents of this chapter is based on the journal paper reporting this very experiment [1]. The reader familiar with our work may recognise the same structure, figures and tables. We wrote an extensive paper about this work and, quite frankly, there is not so much to add here. The only new part is Sect. 2.7, where I devote a few lines to the challenges encountered during this project (mostly FPGA-related) and how they were solved. But the rest of this chapter is a duplicate of our paper [1].

alternative (and more biologically plausible[2]) approach is to progressively adjust the readout weights using various online learning algorithms such as gradient descent [11], recursive least squares [12] or reward-modulated Hebbian learning [13]. Such procedures require minimal data storage and have the advantage of being able to deal with a variable task: should any parameters of the task be altered during the training phase, the reservoir computer would still be able to produce good results by properly adjusting the readout weights. And as will be shown in Chap. 5, online learning allows training complex (and even slightly nonlinear) analogue layers without the challenging task of modelling the underlying structure.

The basic idea of this experiment is to apply this online learning approach to an opto-electronic reservoir computer and show that such an implementation would be well suited for real-time data processing. The use of a FPGA board is inevitable here, as the system needs to be trained in real time, that is, in parallel with the opto-electronic experiment. Such a system could, in principle, be applied to any kind of signal processing tasks, in particular to those that depend on time. A good example of such a task is the wireless channel equalisation, already investigated in previous experiments by our lab (see e.g. [4, 7, 14, 15]), and introduced in Sect. 1.1.4.1. In addition to its potential real-life applications, it can be easily extendable from stationary to time-dependent. This has not been done before, so that is another minor novelty of this experiment. More on that in Sect. 2.2.

Wireless communications is by far the fastest growing segment of the communications industry. The increasing demand for higher bandwidths requires pushing the signal amplifiers close to the saturation point which, in turn, adds significant nonlinear distortions into the channel. These have to be compensated by a digital equaliser on the receiver side [16]. The main bottleneck lies in the Analogue-to-Digital Converters (ADCs) that have to follow the high bandwidth of the channel with sufficient resolution to sample correctly the distorted signal [17]. Current manufacturing techniques allow producing fast ADCs with low resolution, or slow ones with high resolution, obtaining both being very costly. This is where analogue equalisers become interesting, as they could equalise the signal before the ADC and significantly reduce the required resolution of the converters, thus potentially cutting costs and power consumption [18–20]. Moreover, optical devices may outperform digital devices in terms of processing speed [18, 21]. It can for instance be shown that reservoir computing implementations can reach comparable performance to other digital algorithms (namely, the Volterra filter [22]) for equalisation of a nonlinear satellite communication channel [23].

To sum up, the primary goal of this experiment was to investigate the possibility of online training of an opto-electronic RC with a FPGA board. Should this idea work, we then intended to evaluate the performance of the setup on time-dependent

---

[2]Here is a simple example to illustrate the idea: an average language student needs to encounter a new word seven times to memorise it. At each such occurrence, the brain adjusts the connections between neurons, somewhere, and by the seventh time the connection becomes strong enough for the student to quickly remember the not-so-new-anymore word. The same approach can be applied to artificial neural networks, and that is what online learning is all about.

wireless channels. As has been said above, and will be shown in detail in Sect. 2.6, not only we accomplished both parts of the project, but we also discovered unexpected possibilities of the FPGA, that gave birth to the experiment described in Chap. 4.

## 2.2 Equalisation of Non-stationary Channels

The standard version of the channel equalisation task has been introduced in Sect. 1.1.4.1. However, in that version, the task is static, and is thus of little interest for the demonstration of an online-trained system. For that reason, we had to tweak it a little bit. And since the task arises from a real-world problem, making it non-stationary requires almost no effort. One has but to think of the most common every day situations, like emitters and receivers on the move, or obstacles suddenly appearing in the way of the signal. Sections 2.2.2 and 2.2.3 will outline what add-ons we came up with to make the channel equalisation task time-dependent. The next Sect. 2.2.1 discusses an additional study of the parameters of the channel model [24] that we performed in order to better understand its internal mechanism.

### 2.2.1 Influence of Channel Model Parameters on Equaliser Performance

Equations 1.18 and 1.19 model a particular channel with certain amounts of symbol interference and nonlinear distortion, defined by the numerical values of the coefficients employed. To obtain a better understanding of this particular channel model, and to show which stages of input signal distortion are the most difficult to equalise, we introduce a more general channel model, given by

$$
\begin{aligned}
q(n) = {} & (0.08 + m)d(n + 2) - (0.12 + m)d(n + 1) \\
& + d(n) + (0.18 + m)d(n - 1) \\
& - (0.1 + m)d(n - 2) + (0.091 + m)d(n - 3) \\
& - (0.05 + m)d(n - 4) + (0.04 + m)d(n - 5) \\
& + (0.03 + m)d(n - 6) + (0.01 + m)d(n - 7),
\end{aligned} \tag{2.1}
$$

$$
u(n) = p_1 q(n) + p_2 q^2(n) + p_3 q^3(n), \tag{2.2}
$$

and we investigate the equalisation performance for different values of parameters $p_i$ and $m$. To preserve the general shape of the channel impulse response we keep the coefficient of $d(n)$ fixed at 1 in Eq. 2.1. Figure 2.1 shows the resulting impulse responses, given by Eq. 2.1, for several values of $m$. The results of these investigations are presented in Sect. 2.6.5.

**Fig. 2.1**  Various channel impulse responses, given by Eq. 2.1, for different values of $m$. Note that the $d(n)$ coefficient is kept fixed at 1. Dotted curve shows the default shape defined by Eq. 1.18

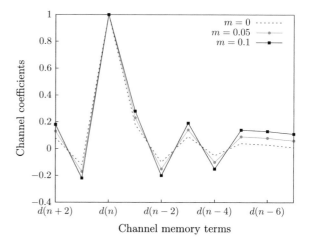

## 2.2.2   Slowly Drifting Channel

The model given by Eqs. 1.18 and 1.19 describes an idealistic stationary noisy wireless communication channel, that is, the channel remains the same during the transmission. However, in wireless communications, the environment has a great impact on the received signal. Given its highly variable nature, the properties of the channel may be subject to important changes in real time.

To investigate this scenario, we performed a series of experiments with a "drifting" channel model, where parameters $p_i$ or $m$ were varying in real time during the signal transmission. These variations occurred at slow rates, much slower than the time required to train the reservoir computer. We studied two variation patterns: a monotonic increase (or decrease) and slow oscillations between two fixed values. Section 2.6.3 shows the results we obtained with our implementation.

## 2.2.3   Switching Channel

In addition to slowly drifting parameters, the channel properties may be subject to abrupt variations due to sudden changes of the environment. For better practical equalisation performance, it is crucial to be able to detect significant channel variations and adjust the RC readout weights in real time. We consider here the case of a "switching" channel, where the channel model switches instantaneously. An example of such a scenario in real life would be, for instance, a cellular phone disconnecting from one base station and connecting to another, with a different signal amplifier. The reservoir computer has to detect such changes and automatically trigger a new training phase, so that the readout weights get adapted for the equalisation of the new channel.

Specifically, instead of a constant channel, given by Eqs. 1.18 and 1.19, we introduce three channels differing in nonlinearity

$$u_1(n) = 1.00q(n) + 0.036q^2(n) - 0.011q^3(n), \tag{2.3a}$$

$$u_2(n) = 0.80q(n) + 0.036q^2(n) - 0.011q^3(n), \tag{2.3b}$$

$$u_3(n) = 0.60q(n) + 0.036q^2(n) - 0.011q^3(n), \tag{2.3c}$$

and switch regularly from one channel to another, keeping Eq. 1.18 unchanged. The results of this experiment are presented in Sect. 2.6.4.

## 2.3 Online Training

The online training approach can be realised through numerous different algorithms. Since we could not try them all, for obvious reasons, we had to pick one for this experiment. Before I got into this project, Anteo Smerieri had already tested in numerical simulations three online training algorithms: simple gradient descent [11], recursive least squares (RLS) [12] and reward-modulated Hebbian learning [13]. The choice was basically dictated by the ease of implementation on the FPGA. While the gradient descent algorithm can be expressed with two very simple equations

$$w_i(n + 1) = w_i(n) + \lambda (d(n) - y(n)) x_i(n),$$
$$\lambda(m + 1) = \lambda_{min} + \gamma (\lambda(m) - \lambda_{min}),$$

where $w_i$ are the readout weights and $\lambda$ is an update rate (we will explain the math in the following section, the equations are given here merely to illustrate the complexity of the algorithms), the RLS algorithm, in its compact version, looks like this

$$k(n) = \frac{\nu^{-1}\Gamma(n - 1)x(n)}{1 + \nu^{-1}x^T\Gamma(n - 1)x(n)},$$
$$w(n) = w(n - 1) + k(n) (d(n) - y(n)),$$
$$\Gamma(n) = \nu^{-1}\Gamma(n - 1) - \nu^{-1}k(n)x^T(n)\Gamma(n - 1),$$

where $k(n), x(n)$ and $w(n)$ are vectors and $\Gamma(n)$ is the estimate of the inverse of the correlation matrix (see Sect. 1.1.3), that is, a $50 \times 50$ matrix for a 50-neuron reservoir. One does not need to understand the precise meaning of these equations to get convinced that the second algorithms is more complex than the simple gradient descent and is, hence, much more complicated to implement in hardware. The two main complications are matrix operations, requiring significant amounts of memory, and division—one of the toughest operations to translate into binary operations (divisions by a power of two being an exception). And despite several non-negligible advantages in performance and convergence speed of RLS, simple gradient descent was the lucky winner. Still, implementing RLS remains an interesting step forward, that would most likely require a certain development period, but the reward will be a much faster convergence to optimal readout weights.

## 2.3.1  Gradient Descent Algorithm

Without further ado, let us introduce the basic principle of the simple gradient descent. Furthermore, in this work we also developed two new variants of the algorithm, that we tested together with the original version. These alternative version are going to be the topic of Sects. 2.3.1.2 and 2.3.1.3.

The gradient, or steepest, descent method is an algorithm for finding a local minimum of a function using its gradient [25]. For the channel equalisation task considered here, the rule for updating the readout weights is given by [11]

$$w_i(n+1) = w_i(n) + \lambda \left( d(n) - y(n) \right) x_i(n), \tag{2.4}$$

where $\lambda$ is the step size, used to control the learning rate. The origin of this procedure is the following: since the error at time $n$ is given by $(d(n) - y(n))^2$ (see Sect. 1.1.3), the derivative of the error with respect to $w_i$ gives $(d(n) - y(n))x_i(n)$, i.e. the right-hand side of Eq. 2.4. At high values of $\lambda$, the weights get close to the optimal values very quickly (in a few steps), but keep oscillating around these values. At low values, the weights converge slowly to the optimal values. In practice, we start with a high value $\lambda = \lambda_0$, and then gradually decrease it during the training phase until a minimum value $\lambda_{min}$ is reached, according to the equation

$$\lambda(m+1) = \lambda_{min} + \gamma \left( \lambda(m) - \lambda_{min} \right), \tag{2.5}$$

with $\lambda(0) = \lambda_0$ and $m = \lfloor n/k \rfloor$, where $\gamma < 1$ is the decay rate and $k$ is the update rate for the parameter $\lambda$.

The gradient descent algorithm suffers from a relatively slow convergence towards the global minimum, but its simplicity, with few simple computational steps, and flexibility, as the convergence rate and the resulting performance can be improved by tuning the parameters $\lambda$ and $\gamma$, make it a reasonable choice for a first implementation on a FPGA chip.

### 2.3.1.1  Full Version

The step size parameter $\lambda$ is used to control the learning rate, and can also be employed to switch the training on or off. That is, setting $\lambda$ to zero stops the training process. This is how experiments on a stationary channel are performed: $\lambda$ is programmed to decay from $\lambda_0$ to 0 during a defined period, and then the reservoir computer performance is tested over a sequence of symbols, with constant readout weights.

#### 2.3.1.2  Non-stationary Version

When equalising a drifting channel, the reservoir should be able to follow the variations and adjust the readout weights accordingly. This can be achieved by setting $\lambda_{min} > 0$ and thus letting the training process continue during the drift of the channel parameters. This procedure was used for experiments described in Sect. 2.6.3.

#### 2.3.1.3  Simplified Version

As mentioned in the previous paragraph, the equalisation of a non-stationary channel requires keeping $\lambda_{min} > 0$. However, this worsens the equalisation performance, as the readout weights keep oscillating around the optimal values. This can be seen from Eq. 2.4, that defines the update rule for the readout weights: at each time step $n$, a small correction $\Delta w_i = \lambda(n)(d(n) - y(n))x_i(n)$ is added to every weight $w_i$. These corrections are gradually reduced by decreasing the learning rate $\lambda(n)$, so that the weights converge to their asymptotic values. In the case of a constant $\lambda$, the corrections $\Delta w_i$ are only damped by the error $d(n) - y(n)$, which stops decreasing at some point, leaving the $w_i$ oscillating around the optimal values.

To check the impact of a constant $\lambda$ on the equalisation performance we performed several experiments with a simplified version of the training algorithm by setting $\gamma = 0$, and hence $\lambda(n) = \lambda_0$ for all $n$. Although this method will increase the error slightly, it has several advantages. With $\lambda$ constant, there is no need to search for an optimal decay rate $k$, which results in fewer experimental parameters to scan and thus shorter overall experiment runtime. Keeping $\lambda$ at a constant, non-zero value would also allow the equaliser to follow a drifting channel, as described in Sect. 2.2.2. The results obtained with this simplified version of the algorithm are shown in Sect. 2.6.2.

### 2.4  Experimental Setup

Our experimental setup is depicted in Fig. 2.2. It contains three distinctive components: the opto-electronic reservoir, the FPGA board implementing the input and the readout layers and the computer used to setup the devices and record the results. The reader should already be familiar with the reservoir part—it has been thoroughly discussed in Sect. 1.2.4. Thus, in the following sections we will focus on the new components: the FPGA board (Sect. 2.4.1) and the computer (Sect. 2.4.3). Additionally, Sect. 2.4.2 outlines the experimental parameters, tuned to obtain the best results.

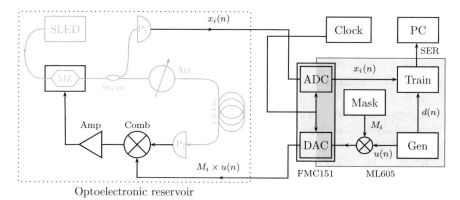

Optoelectronic reservoir

**Fig. 2.2** Schematic representation of the experimental setup. The opto-electronic reservoir has been introduced in Sect. 1.2.4. The FPGA board implements both the input and output layers, generating the input symbols and training the readout weights. The computer controls the devices and records the results

## 2.4.1   Input and Readout

For our implementation, we use the Xilinx ML605 evaluation board (see Fig. 2.3), powered by the Virtex 6 XC6VLX240T FPGA chip. The board is equipped with a JTAG port, used to load the FPGA design onto the chip, and a UART port, that we use to communicate with the board (as described in Sect. 2.5). The LPC (Low Pin Count) FMC (FPGA Mezzanine Card) connector is used to attach the 4DSP FMC151 daughter card, containing one two-channel ADC (Analogue-to-Digital converter) and one two-channel DAC (Digital-to-Analogue converter). The ADC's maximum sampling frequency is 250 MHz with 14-bit resolution, while the DAC can sample at up to 800 MHz with 16-bit precision.

The synchronisation of the FPGA board with the reservoir delay loop is crucial for the performance of the experiment. For proper acquisition of reservoir states, the ADC has to output an integer number of samples per roundtrip time. The daughter card contains a flexible clock tree, that can drive the converters either from the internal clock source, or an external clock signal. As the former is limited to the fixed frequencies of the onboard oscillator, we employ the latter option. The clock signal is generated by a Hewlett Packard 8648 A signal generator. With a reservoir of $N = 51$ neurons (one neuron is added to desynchronise the inputs from the reservoir, as has been discussed in Sect. 1.2.3) and a roundtrip time of 7.94 μs, the sampling frequency is set to 128.4635 MHz, thus producing 20 samples per reservoir state. To get rid of the transients, induced mainly by the finite bandwidths of the ADC and DAC, the 6 first and 6 last samples are discarded, and the neuron value is averaged over the remaining 8 samples.

The potentials of the electric signals to and from the mezzanine card need to be adjusted in order to achieve the most efficient interface without damaging the

**Fig. 2.3** Xilinx ML605 board with Virtex 6 FPGA chip and 4DSP FMC150 daughter card (FMC150 and FMC151 cards look practically the same). Image reprinted with permission from www.fpgadeveloper.com

hardware. The DAC output voltage of 2 $V_{p-p}$ is sufficient for this experiment, as typical voltages of the input signal range between 100 mV and 200 mV. The ADC is also limited to 2 $V_{p-p}$ input voltage. With settings described in the previous section, the output voltage of the readout photodiode does not exceed 1 $V_{p-p}$.

## 2.4.2  Experimental Parameters

To achieve the best performance, we scan the most influential parameters, which are: the input gain $\beta$, the decay rate $k$, the channel signal-to-noise ratio and the feedback attenuation, that corresponds to the feedback gain parameter $\alpha$ in Eq. 1.5. The first three parameters are set on the FPGA board, while the last one is tuned on the optical attenuator. The input gain $\beta$ is stored as a 18-bit precision real in [0, 1[ and was scanned in the [0.1, 0.3] interval. The decay rate $k$ is an integer, typically scanned from 10 up to 50 in a few wide steps. The noise ratios were set to several pre-defined values, in order to compare our results with previous reports. The feedback attenuation was scanned finely between 4.5 dB and 6 dB. Lower values would allow cavity oscillations to disturb the reservoir states, while higher values would not provide enough feedback to the reservoir. Table 2.1 contains the values of parameters we used for the gradient descent algorithm (defined in Sect. 2.3.1).

**Table 2.1** Gradient descent algorithm parameters

| $\lambda_0$ | $\lambda_{min}$ | $\gamma$ | $k$ |
|---|---|---|---|
| 0.4 | 0 | 0.999 | 10–50 |

### *2.4.3  Experiment Automation*

The experiment is fully automated and controlled by a Matlab script, running on a computer. It is designed to run the experiment multiple times over a set of predefined values of parameters of interest and select the combination that yields the best results. For statistical purposes, each set of parameters is tested several times with different random input masks (see Sect. 1.1.3).

At launch, connections to the optical attenuator and the FPGA board are established, and the parameters on the devices are set to default values. After generating a set of random input masks, the experiment is run once and the elapsed time is measured. The duration of one run depends on the lengths of train and test sequences and varies from 6 s to 12 s. This is considerably shorter than the offline-trained implementation [4], that required about 30 s. The script runs through all combinations of scanned parameters. For each combination, the values of the parameters are sent to the devices, the experiment is run several times with different input masks and the resulting error rates (see Sect. 2.5) are stored in the Matlab workspace. Once all the combinations are tested, the connections to the devices are closed and all collected data is saved to a file.

## 2.5  FPGA Design

Rejoice, here comes my favourite section of this chapter, that describes the FPGA design—the part of the experiment on which I spent the majority of the time. Ironically, most readers will probably want to skip, or merely overlook this section, since it is, I admit, quite technical and not so crucial for understanding of the results in Sect. 2.6. But I could not omit it, since this is the heart of my work, the part I put most energy and time in. Therefore, I welcome any interested reader to venture in the following paragraphs, but remind once again that you will not miss any key points if you just go ahead to the next section.

The FPGA design is written in standard IEEE 1076-1993 VHDL language [19, 26] and compiled with Xilinx ISE Design Suite 14.7, provided with the board. We also used Xilinx ChipScope Pro Analyser to monitor signals on the board, mostly for debugging and testing.

The simplified schematics of our design is depicted in Fig. 2.4. Rectangular boxes represent modules (i.e. entities) and the lines stand for data connections between them. As discussed in Sect. 2.4.1, the FPGA board implements both the input and the readout layers of the reservoir computer. The board has a digital connection to a computer (running a Matlab script) and an analogue one to the experimental setup. The former, realised through a UART port bridged to a standard COM port, is used to load parameters (e.g. $\lambda_0, \gamma, \ldots$) into the board and read the experiment results (i.e. symbol error rate) from the board. The latter consists of three analogue connections: an output signal to the reservoir, containing the masked inputs

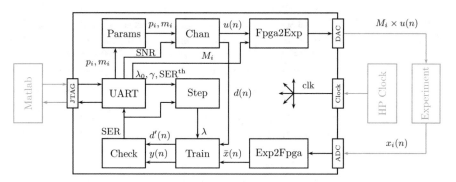

**Fig. 2.4** Simplified schematics of the FPGA design. Smaller boxes and arrows inside the board (large black box) represent modules (entities) and signals. The input layer modules (Params, Chan, and Fpga2Exp) generate the target signal $d(n)$ and compute a nonlinear channel output $u(n)$. The readout layer (Exp2Fpga, Train, and Step) receives the reservoir states $x_i(n)$ from the experiment, trains the weights $w_i$ and computes the output signal $y(n)$. The Check module evaluates the symbol error rate. The UART module executes commands issued by Matlab, sets variable parameters and sends the results back to the computer

$M_i \times u(n)$, a clock signal clk from the HP signal generator and an input signal from the readout photodiode, containing reservoir states $x_i(n)$.

The operation of the FPGA board is controlled from the computer. A predefined set of 4-byte commands can be transmitted through the JTAG port, such as write a specific parameter value into the appropriate register or toggle the board state from reset to running, and vice versa. The commands are received and executed by the UART module. In addition, when the FPGA is running, the module regularly transmits the value of the SER signal to the computer. In order to prevent collisions in the UART channel, commands from computer are only sent when the board is in a reset state, that is, no channel is being equalised.

The Chan module implements the nonlinear channel model, given by Eqs. 1.18 and 1.19, and generates the input signal for the reservoir. It receives the noise amplitude, for a defined Signal-To-Noise ratio, from the computer via UART module. The channel parameters $p_i$ and $m_i$ are supplied by the Params module. Two Galois Linear Feedback Shift Registers (GLFSRs) with a total period of about $10^9$ are used to generate pseudorandom symbols $d(n) \in \{-3, -1, 1, 3\}$. Another GLFSR of period around $2 \times 10^5$ generates pseudorandom numbers used as noise $\nu(n)$. The symbol sequence $d(n)$ is sent to the Train module as a target signal, while the channel output $u(n)$ is multiplied by the input mask $M_i$ within the Fpga2Exp module, and then converted to an analogue signal by the FMC151 daughter card.

The analogue reservoir output $x_i(n)$ is converted into a digital signal by the ADC. The time-multiplexed reservoir states are then sampled and averaged by the Exp2Fpga module, which transmits all the neurons from one reservoir $\bar{x}(n)$ in parallel to the next module.

The synchronisation of the readout layer with the opto-electronic reservoir is performed by both Fpga2Exp and Exp2Fpga modules. At the beginning of a run of the experiment, the former sends a short pulse into the reservoir, before transmitting the input symbols. This pulse is detected by the Exp2Fpga module and then used to synchronise the sampling and averaging process with the incoming reservoir states.

The Train module implements the simple gradient descent algorithm. It receives the neurons $\bar{x}(n)$, the target signal $d(n)$ and the gradient step $\lambda$, computes the reservoir output $y(n)$ with its error from the target signal, and adjusts the readout weights $w_i$ following Eq. 2.4. The input target signal $d(n)$ is delayed by several periods $T$ to compensate the propagation time of the information through the input layer, the opto-electronic reservoir and the Exp2Fpga module. The reservoir output $y(n)$ is then rounded up to the closest channel symbol $y(n) \rightsquigarrow \{-3, -1, 1, 3\}$ and compared to the delayed target signal $d'(n)$ by the Check module, that counts misclassified symbols and outputs the resulting Symbol Error Rate.

The evolution of the learning rate $\lambda$ is governed by a separate module Step, which implements the Eq. 2.5, with initial value $\lambda_0$ and decay rate $\gamma$ set on the computer and transferred to the board through the UART connection. The module also monitors the performance of the reservoir computer and resets $\lambda$ to its initial value $\lambda_0$ when the Symbol Error Rate exceeds a predefined threshold value $SER^{th}$. This feature is used for the switching channel (see Sects. 2.2.3 and 2.6.4) and allows to improve the performance of the system by adjusting the readout weights to the new channel parameters.

The gradient descent algorithm is relatively simple, with only few addition and multiplication operations involved in Eqs. 2.4 and 2.5. While an adder can easily be built with a small amount of logic gates, multiplication is more complicated to implement and requires lots of resources. Moreover, as all readout weights are computed in parallel, the size of the design grows quickly with the number of neurons $N$. This results in slow implementation process and very low chances of generating a design that functions correctly. The solution resides in the use of special DSP48E slices, designed and optimised to perform a predefined set of arithmetic operations [28]. With proper settings, this dedicated microprocessor is capable of performing a 25 bit $\times$ 18 bit multiplication in less than 6 ns. While the speed gain compared to standard logic blocks is minimal, the implementation of the FPGA design is greatly simplified, as hundreds of logic gates and registers get replaced by just one component.

The arithmetic operations mentioned above are performed on real numbers. However, a FPGA is a logic device, designed to operate with bits. The performance of the design thus highly depends on the bit-representation of real numbers, i.e. the precision. The main limitation comes from the DSP48E slices, as these are designed to multiply a 25-bit integer by another 18-bit integer. To meet these requirements, our design uses a fixed-point representation with different bit array lengths for different variables. Parameters and signals that stay within the $]-1, 1[$ interval are represented by 18-bit vectors, with 1 bit for the sign and 17 for the decimal part. These are the learning algorithm parameters $\lambda$, $\lambda_0$ and $\gamma$, the input mask elements $M_i$ and the reservoir states $x_i(n)$, extended from the 14-bit ADC output. Other variables, such as reservoir output $y(n)$ and readout weights $w_i$ span a wider $[-16,16]$ interval and

**Table 2.2** Total usage of FPGA resources

|             | Registers | LUTs   | Block RAM | DSP48E |
|-------------|-----------|--------|-----------|--------|
| Used        | 12288     | 5661   | 198       | 161    |
| Available   | 301440    | 150720 | 416       | 768    |
| Utilisation | 4%        | 3%     | 47%       | 20%    |

are represented as 25-bit vectors, with 1 sign bit, 4 bits for the integer part and 20 bits for the decimal part.

Table 2.2 reports total FPGA resource usage of our implementation. The design requires relatively few registers and Lookup Tables (LUTs). Most of the arithmetic operations are performed by the DSP48E slices, and their number grows roughly as $3 \times N$, thus theoretically limiting our reservoir to 255 neurons. Note that this restriction can be easily overcome by rearranging the DSP48E slices in a less concurrent design. High internal memory (block RAM) usage is due to several ChipScope modules (not shown in Fig. 2.4), added to monitor internal FPGA signals. To conclude, our implementation can be expanded to work with much bigger reservoirs.

## 2.6 Results

This section presents the results of different investigations outlined in Sects. 2.2 and 2.3.1. All results presented here were obtained with the experimental setup described in Sect. 2.4.

### 2.6.1 Improved Equalisation Error Rate

Figure 2.5 presents the performance of our reservoir computer for different Signal-to-Noise Ratios (SNRs) of the wireless channel (black line). We investigated realistic SNR values for real world channels such as 60 GHz LAN [29] and Wi-Fi [30]. For each SNR, the experiment was repeated 20 times with different random input masks. Average SERs are plotted on the graph, with error bars corresponding to maximal and minimal values obtained with particular masks. We used noise ratios from 12 dB up to 32 dB, and also tested the performance on a noiseless channel, that is, with infinite SNR. The RC performance was tested over one million symbols, and in the case of a noiseless channel the equaliser made zero error over the whole test sequence with most input masks.

The experimental parameters, such as the input gain $\beta$ and the feedback attenuation $\alpha$, were optimised independently for each input mask. Figure 2.6 shows the dependence of the SER on these parameters. The plotted SER values are averaged

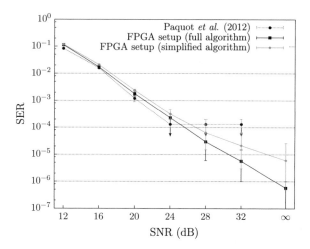

**Fig. 2.5** Experimental results obtained with our setup. Symbol Error Rates (SERs) are plotted against the Symbol-to-Noise Ratio (SNR). The equaliser was tested with 20 different random input masks over one million input symbols, average values are plotted on the graph (black line). For the noiseless channel (SNR = ∞), for most choices of input mask, the RC made no errors over the test sequence. The dotted line shows the results of the opto-electronic setup with offline training [4]. For low noise levels, our system produces error rates significantly lower than [4], and for noisy channels the results are similar. The gray line depicts the SERs obtained with the simplified version of the training algorithm (see Sect. 2.3.1.3). The equalisation is less efficient than with the full algorithm, but the optimisation of experimental parameters takes less time

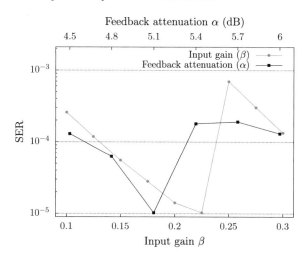

**Fig. 2.6** Dependence of the equaliser performance (at 32 dB SNR) on the experimental parameters. Average SERs (over 10 random input masks) are plotted against the input gain (gray line) and the feedback attenuation (black line). The optimal feedback attenuation has to be set around $5.1 \pm 0.3$ dB, outside this region the SER deteriorates by roughly one order of magnitude. The input gain shows a minimum around $0.225 \pm 0.025$

**Table 2.3** Optimal reservoir computer parameters

| $\alpha$ | $\beta$ | $V_\phi$ |
|---|---|---|
| $5.1 \pm 0.3$ dB | $0.225 \pm 0.025$ | 1.6 V |

over 10 random input masks. For this figure, we used data from a different experiment run with more scanned values. For each curve, the non-scanned parameter was set to the optimal value. The equaliser shows moderate dependence on both parameters, with an optimal input gain located within $0.225 \pm 0.025$ and an optimal feedback attenuation of $5.1 \pm 0.3$ dB. The MZ modulator DC bias voltage $V_\phi$ is set to 1.6 V, which yields a slightly shifted transfer function in order to compensate the input symbols distribution (see Sect. 1.1.4.1). These parameters are summarised in Table 2.3.

We compare our results to those reported in [4], obtained with the same optoelectronic reservoir, trained offline (dottes line). For high noise levels (SNR $\leq$ 20 dB) our results are similar to those in [4]. For low noise levels (SNR $\geq$ 24 dB) the performance of our implementation is significantly better. Note that the previously reported results are only rough estimations of the equaliser's performance as the input sequence was limited by hardware to 6000 symbols [4]. In our experiment the SER is estimated more precisely over one million input symbols. For the lowest noise level (SER = 32 dB) an SER of $1.3 \times 10^{-4}$ was reported in [4], while we obtained an error rate of $5.71 \times 10^{-6}$ with our setup. One should remember that common error detection schemes, used in real-life applications, require the SER to be lower than $10^{-3}$ in order to be efficient. Digital equalisation algorithms, based on bilinear decision feedback equalisers [24] and ESNs with random reservoirs [31], report error rates of $10^{-3}$ and $10^{-5}$, respectively. These values are only provided for illustration, as it makes little sense to compare numerical algorithms and physical systems.

To the best of our knowledge, the results presented here (at 32 dB SNR) are the lowest error rates ever obtained with a standalone experimental reservoir computer. SERs around $10^{-4}$ have been reported in [4, 66, 67] and a passive cavity based setup [14] achieved a $1.66 \times 10^{-5}$ rate (this values is limited by the use of a 60000-symbol test sequence), but no results below $10^{-5}$ have been published so far. However, this is not the main achievement of this experiment. Indeed, had it been possible to test [4] on a longer sequence, it is possible that comparable SERs would have been obtained. The strength of this setup resides in the adaptability to changing environment, as will be shown in the following sections.

## 2.6.2  Simplified Training Algorithm

The performance of the simplified training algorithm is shown in Fig. 2.5 (gray line). The equaliser was tested with 10 random input masks and one million input symbols, the training was performed over 100000 symbols. Only three parameters were scanned during these experiments: the input gain $\beta$, the feedback attenuation $\alpha$ and the signal-to-noise ratio. The learning rate $\lambda$ was set to 0.01. The overall experimental runtime was significantly shorter: while an experiment with full training algorithm would last for about 50 h, these results were obtain in approximately 10 h (which is due to five different values of $k$ tested in the former case).

For high noise levels the results of the two algorithms are close and for low noise levels the simplified version yields slightly worse error rates. The performance is much worse in the noiseless case and strongly depends on the input mask: we notice a difference of almost two orders of magnitude between the best and the worst result. This performance loss is the price to pay for the simplified algorithm and shorter experimental runtime.

### 2.6.3  Equalisation of a Slowly Drifting Channel

Besides the environmental conditions, the relative positions of the emitter and the receiver can have a significant impact on the properties of a wireless channel. A simple example is a receiver moving away from the transmitter, causing the channel to drift more or less slowly, depending on the relative speed of the receiver. Here we show that our Reservoir Computer is capable of dealing with drifts with time scales of order of a second. This time scale is in fact slow compared to those expected in real life situations, but the setup could be sped up by several orders of magnitude, as will be shown in the next section.

A drifting channel is a good example of a situation where training the reservoir online yields better results than offline. We have previously shown in numerical simulations that training a reservoir computer offline on a non-stationary channel results in an error rate ten times worse than with online training [32]. We demonstrate here that an online-trained experimental reservoir computer performs well even on a drifting channel if $\lambda_{min}$ is set to a small non-zero value (see Sect. 2.3.1.2).

At first, we investigated the relationship between the channel model coefficients and the lowest error rate achievable with our setup. That is, would the equalisation performance be better or worse if one of the numerical values in Eqs. 1.18 and 1.19 was changed by, for instance, 10%. Given the vast amount of possibilities of varying the 4 parameters $p_i$ and $m$, we picked those that seemed most interesting and most significant. We thus tested the amplitude of the linear part, given by the parameter $p_1$, the amplitude of the quadratic and cubic parts, given by $p_2$ and $p_3$, and the memory $m$ of the impulse response. For each test, only one aspect of the channel was varied and other parameters were set to default values (as in Eqs. 1.18 and 1.19). The results of these investigations are presented in Sect. 2.6.5.

We then programmed these parameters to vary during experiments in two different ways: a monotonic growth (or decay) and a periodic linear oscillation between two defined values. The results of these experiments are depicted in Fig. 2.7.

Figure 2.7A shows the experimental results for the case of monotonically decreasing $p_1$ from 1 to 0.652. The solid gray curve presents the resulting SER with $\lambda_{min} = 0$, that is, with training process stopped after 45000 input symbols. The solid black curve depicts the error rate obtained with $\lambda_{min} = 0.01$, so that the readout weight can be gradually adjusted as the channel drifts. Note that while in the first experiment the SER grows up to 0.329, it remains much lower in the second case. The increasing error rate in the latter case is due to the decrease of $p_1$ resulting in a more complex channel. Dashed black curves show the best possible error rate obtained with our setup for different values of $p_1$, as presented in Sect. 2.6.5. With $p_1$ approaching 0.652, the obtained error rate is $8.0 \times 10^{-3}$, which is the lowest error rate achievable for this value of $p_1$, as demonstrated in Fig. 2.9a. This shows that the non-stationary version of the training algorithm allows a drifting channel to be equalised with the lowest error rate possible.

Figure 2.7B depicts error rates obtained with $p_1$ linearly oscillating between 1 and 0.688. With $\lambda_{min} = 0$ (solid gray curve) the error rate is as low as $1 \times 10^{-4}$ when $p_1$ is around 1, and grows very high elsewhere. With $\lambda_{min} = 0.01$, the obtained SER is always at the lowest value possible: at the point where $p_1 = 0.688$, it stays at $5.0 \times 10^{-3}$, which again is close to the best performance for such channel, illustrated by the solid black curve.

We obtained similar results with parameters $p_2$, $p_3$ and $m$, as shown in Figs. 2.7C–G. Letting the reservoir computer adapt the readout weights by setting $\lambda_{min} > 0$ produces the lowest error rates possible for a given channel, while stopping the training with $\lambda_{min} = 0$ results in quickly growing SERs.

## 2.6.4 Equalisation of a Switching Channel

Figure 2.8 shows the error rate produced by our experiment in case of a switching noiseless communication channel. The parameters of the channel are programmed to switch in cycle among Eq. 2.3 every 266000 symbols. Every switch is followed by a steep increase of the SER, as the reservoir computer is no longer optimised for the channel it is equalising. The performance degradation is detected by the algorithm, causing the learning rate $\lambda$ to be reset to the initial value $\lambda_0$, and the readout weights are re-trained to new optimal values.

For each value of $p_1$, the reservoir computer is trained over 45000 symbols, then its performance is evaluated over the remaining 221000 symbols. In case of $p_1 = 1$, the average SER is $1 \times 10^{-5}$, which is the expected result. For $p_1 = 0.8$ and $p_1 = 0.6$ we compute average SERs of $7.1 \times 10^{-4}$ and $1.3 \times 10^{-2}$, respectively, which are the best results achievable with such values of $p_1$ according to our previous investigations (see Fig. 2.9a). This shows that after each switch the readout weights are updated to new optimal values, producing the best error rate for the given channel.

Note that the current setup is rather slow for practical applications. With a roundtrip time of $T = 7.94\,\mu s$, its bandwidth is limited to 126 kHz and training the reservoir over 45k samples requires 0.36 s to complete. However, it demonstrates the potential of such systems in equalisation of non-stationary channels. For

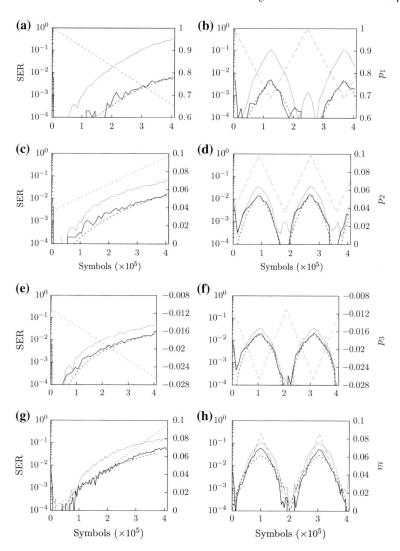

**Fig. 2.7** Symbol error rates (left axis, log scale), averaged over 10 k symbols, produced by the experimental setup with a drifting channel. Each panel presents data obtained from one experiment run with a fixed input mask and optimal parameters $\alpha$, $\beta$ and $k$. Two different training methods were tested: solid gray curves show the results produced by the full training algorithm with $\lambda_{min} = 0$ (see Sect. 2.3.1.1), while solid black curves depict those obtained with the non-stationary version with $\lambda_{min} > 0$ (see Sect. 2.3.1.2). Dashed black lines display the best performance that can be obtained with our system (the details are discussed in Sect. 2.6.5) for given values of variable parameters $p_i$ and $m$ (right axis, linear scale), shown in dashed gray (see Sect. 2.6.5 for details). The $x$-axis, shown in symbols, can also be expressed as time, given that one symbol is processed in $T = 7.94\,\mu s$. **a b** Monotonically decreasing and oscillating $p_1$. **c & d** Monotonically increasing and oscillating $p_2$. Figure continued on next page. **d & e** Monotonically decreasing and oscillating $p_3$. **f & g** Monotonically increasing and oscillating $m$

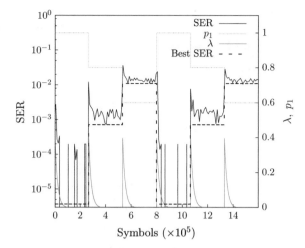

**Fig. 2.8** Symbol error rate (left axis, solid black line), averaged over 10000 symbols, produced by the FPGA in case of a switching channel. The value of $p_1$ (right axis, dotted gray line) is modified every 266000 symbols. The change in channel is followed immediately by a steep increase of the SER. The $\lambda$ parameter (right axis, solid gray line) is automatically reset to $\lambda_0 = 0.4$ every time a performance degradation is detected, and then returns to its minimum value, as the equaliser adjusts to the new channel, bringing down the SER to its asymptotic value. After each variation of $p_1$, the reservoir re-trains. The lowest error rate possible for the given channel is shown with the dashed black curve

real-life applications, such as for instance Wi-Fi 802.11g, a bandwidth of 20 MHz would be required. This could be realised with a 15 m fibre loop, thus resulting in a delay of $T = 50$ ns. This would also decrease the training time down to 2.2 ms and make the equaliser more suitable for realistic channel drifts. The speed limit of our setup is set by the bandwidth of the different components, and in particular of the ADC and DAC. For instance with $T = 50$ ns and keeping $N = 50$, reservoir states should have a duration of 1 ns, and hence the ADC and DAC should have bandwidths significantly above 1 GHz (such performance is readily available commercially). As an illustration of how a fast system would operate, we refer to the optical experiment [8] in which information was injected into a reservoir at rates beyond 1 GHz.

## 2.6.5 Influence of Channel Model Parameters on Equaliser Performance

This section presents the results obtained after in-depth investigation of the channel model (Eqs. 1.18 and 1.19) and its parameters, as discussed in Sect. 2.2.1.

Figure 2.9a shows the equalisation results for different values of $p_1$. We tested each value over 10 random input masks, with independent experimental parameters optimisation for each run. Average values are presented on the plot, with error bars

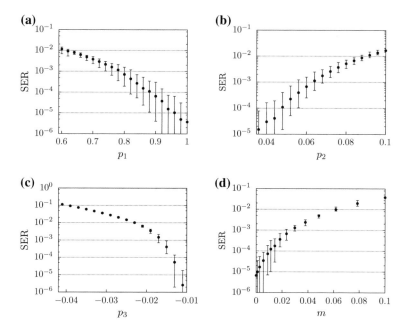

**Fig. 2.9** Error rates for different values of channel parameters $p_i$ and $m$. The results were measured over one million input symbols, with 10 random input masks and zero noise. **a** Lower $p_1$ implies lower linear part of the channel, containing the useful signal, which naturally results in higher error rates. **b** Increasing the quadratic component $p_2$ of the channel makes it more nonlinear, and thus more difficult to equalise. **c** Increasing the cubic component $|p_3|$ of the channel makes it more nonlinear, and thus more difficult to equalise. **d** Higher values of $m$ make the channel equalisation more complex

depicting best and worst results obtained among different masks. The equaliser performance was tested on a sequence of one million inputs, and in several cases we obtained zero misclassified symbols. Note that the observed increase of the SER with reduction of $p_1$ is natural as the linear part contains the signal to be extracted. When decreasing $p_1$, not only the useful signal gets weaker, but the nonlinear distortion also becomes relatively more important.

Figures 2.9b, c present the dependence of the SER on parameters $p_2$ and $p_3$, respectively. These parameters define the amplitude of the nonlinear distortion of the signal, and as they grow, the channel becomes more nonlinear and thus more difficult to equalise. The results of equalisation with different values of $m$ are shown in Fig. 2.9d, higher values of $m$ increase the temporal symbol mixing of the channel, hence worse results.

## 2.7 Challenges and Solutions

The previous Sect. 2.6 listed all the successful results we obtained with this project. The present section is devoted to the other side of the coin, that is, the challenges and problems we encountered in this work. Such a presentation is of little interest for a scientific paper, where only final results matter, but could interest a reader trying to replicate this experiment, partially or in full, who seeks to find out how we actually managed to make the setup work. And since the real new part of this experiment if the FPGA board (the opto-electronic reservoir was already a well-known and functioning system when I started this project), the FPGA design was the main source of troubles. Here they are, and how I managed to solve them.

The first thing that comes to mind is the very steep learning curve of FPGA programming and a completely wrong approach adopted by myself, mostly because of lack of knowledge. Unlike most other programming languages, learning as you go is a very bad idea. There is so much to learn and understand about FPGA internals and mechanisms before even thinking of going somewhere. Well, certainly, very basic, academic designs could be completed while learning, but not a complex one, such as the one described in Sect. 2.5, involving non-trivial logic, several clock domains, multi-cycle paths and complex timing constraints. This is, probably, the reason for such a slow start of my thesis.

Completing a FPGA design is a very long task. The implementation time, that is, the compilation process, from the VHDL code all the way down to the FPGA bitstream, of a very basic design takes a couple of minutes on my high-end laptop. A medium design would require from 30 min to an hour. Complex designs, approaching 80–90% of total resources utilisation of a large FPGA would take hours, or even a day to compile. Fortunately, the above design falls in the second category, with a compilation time of approximately 40 min. Imagine the oh-so-common debugging process, with an error occurring at hardware level, where the simulator can no longer help. Make a small correction, launch the compilation and go for lunch. Then, as the correction did not help, try something else, launch the compilation and…go for lunch again? To sum up, at the latest stages of design development, the compilation times is the process that takes the bulk of time.

Programming usually comes with errors—humans have this sad tendency. The most harmful and deceptive problem I have encountered during the development of this project, that caused a few broken objects and lots of facial hair torn, is the well-known overflow. All numbers in the design are allocated a strict number of bits (usually 18 or 25, see Sect. 2.5) and if the scale is chosen poorly, a summation or multiplication of these numbers may result in a number too big for such an amount of bits. Which results in the exceeding part being cut off, and the calculation is performed normally. But the result is, obviously, incorrect. And such errors are not easy to detect. Moreover, such an overflow may occur outside the trial case, that is, when the design has to deal with numbers it has not been tested on. In other words, the designer thinks that the code works, but, in fact, it does not. Such unexpected overflow was the reason of relatively poor equalisation performance of the experiment, published in a preliminary conference paper [32]. The reservoir was actually equalising a more complex channel, and some of the symbols in the training and test sequences were simply wrong. Once this problem was discovered and corrected, we could obtain the results shown in Fig. 2.5.

Lastly, let me say a few words about the DSP48 slices. At first, I hated them— Xilinx software infers them automatically by default and activates all their internal registers, that cause a mess in the timing of the design. I was very pleased when I found the option to disable them. All multiplications were performed using slice logic at that stage and the compilation was taking a few hours for a 50-neuron design. This strategy did not seem viable, so I invested some time into understanding the DSP48 slices. Now, I love them. They are very complex, but they have so much to offer! And instead of using hundreds of logic gates and registers, only one DSP slice is required to perform a multiplication. Much to my joy, the implementation time was greatly reduced.

## 2.8   Conclusion

In this chapter we covered the first experiment realised for my Ph.D. In this work we applied the online learning approach to training an opto-electronic reservoir computer. We programmed the simple gradient descent algorithm on a FPGA chip and tested our system on the nonlinear channel equalisation task. We obtained error rates up to two orders of magnitude lower than previously reported RC implementations on the channel equalisation task, while significantly reducing the experimental runtime. We also demonstrated that our system is well-suited for non-stationary tasks by equalising a drifting and a switching channel. In both cases, we obtained the lowest error rates possible with our setup.

So far, so good; what's next? From our team's point of view, this experiment was a first step—we employed a FPGA board for the first time and wanted to see what it was capable of and whether it was worth investigating this direction further. And this work has shown that, in fact, it definitely is. Using the FPGA to drive the opto-electronic reservoir allows to perform more intricate operations on the experiment, such as compute the output in real time or, as will be shown in Chap. 4, feed the output back as input. This, in turn, allows to investigate several novel features. These findings greatly influenced the further plans for my thesis, as I was focused on implementing these novel features.

In our journal paper [1], we proposed two ideas for future work. At the moment of writing these lines, we have realised them both. First, a FPGA-driven photonic reservoir computer could implement a full optimisation of the readout weights and the input mask, as suggested in [33, 34]. With Michiel Hermans, we performed that experiment right afterwards. We will talk about it in Chap. 3. Second, the real-time training makes it possible to feed the output signal back into the reservoir. This additional feedback would highly enrich the dynamics of the system, allowing one to tackle new tasks such as pattern generation or chaotic series prediction [35]. And we have demonstrated that experimentally in early 2016. Chapter 4 is devoted to this work.

To conclude, I believe that this experiment is the key achievement of my Ph.D, that made all subsequent results much easier. Therefore, it was two years well spent.

# References

1. Antonik, Piotr, François Duport, Michiel Hermans, Anteo Smerieri, Marc Haelterman, and Serge Massar. 2017. Online training of an opto-electronic reservoir computer applied to real-time channel equalization. *IEEE Transactions on Neural Networks and Learning Systems* 28 (11): 2686–2698.
2. Vandoorne, Kristof, Wouter Dierckx, Benjamin Schrauwen, David Verstreten, Roel Baets, Peter Bienstman, and Jan Van Campenhout. 2008. Toward optical signal processing using photonic reservoir computing. *Optics Express* 16: 11182–11192.
3. Appeltant, Lennert, Miguel Cornelles Soriano, Guy Van der Sande, Jan Danckaert, Serge Massar, Joni Dambre, Benjamin Schrauwen, Claudio R Mirasso, and Ingo Fischer. 2011. Information processing using a single dynamical node as complex system. *Nature Communications* 2: 468.
4. Paquot, Yvan, François Duport, Anteo Smerieri, Joni Dambre, Benjamin Schrauwen, Marc Haelterman, and Serge Massar. 2012. Optoelectronic reservoir computing. *Scientific Reports* 2: 287.
5. Larger, Laurent, M.C. Soriano, Daniel Brunner, L. Appeltant, Jose M. Gutiérrez, Luis Pesquera, Claudio R. Mirasso, and Ingo Fischer. 2012. Photonic informtion processing beyond Turing: an optoelectronic implementation of reservoir computing. *Optics Express* 20: 3241–3249.
6. Martinenghi, Romain, Sergei Rybalko, Maxime Jacquot, Yanne Kouomou Chembo, and Laurent Larger. 2012. Photonic nonlinear transient computing with multiple-delay wavelength dynamics. *Physical Review Letters* 108: 244101.
7. Duport, François, Bendix Schneider, Anteo Smerieri, Marc Haelterman, and Serge Massar. 2012. All-optical reservoir computing. *Optics Express* 20: 22783–22795.
8. Brunner, Daniel, Miguel C Soriano, Claudio R Mirasso, and Ingo Fischer. 2013. Parallel photonic information processing at gigabyte per second data rates using transient states. *Nature Communications* 4: 1364
9. Dejonckheere, Antoine, François Duport, Anteo Smerieri, Li Fang, Jean-Louis Oudar, Marc Haelterman, and Serge Massar. 2014. All-optical reservoir computer based on saturation of absorption. *Optics Express* 22: 10868–10881.
10. Vandoorne, Kristof, Pauline Mechet, Thomas Van Vaerenbergh, Martin Fiers, Geert Morthier, David Verstraeten, Benjamin Schrauwen, Joni Dambre, and Peter Bienstman. Experimental demonstration of reservoir computing on a silicon photonics chip. *Nature Communications* 5: 3541.
11. Bishop, Christopher M. 2006. *Pattern recognition and machine learning*. Springer
12. Haykin, Simon. 2000. *Adaptive filter theory*. Upper Saddle River, New Jersey: Prentice-Hall.
13. Bottou, Léon. 1998. *Online algorithms and stochastic approximations. Online Learning and Neural Networks*. Cambridge University Press
14. Vinckier, Quentin, Françcois Duport, Anteo Smerieri, Kristof Vandoorne, Peter Bienstman, Marc Haelterman, and Serge Massar. 2015. High-performance photonic reservoir computer based on a coherently driven passive cavity. *Optica* 2 (5): 438–446.
15. Duport, François, Anteo Smerieri, Akram Akrout, Marc Haelterman, and Serge Massar. 2016. Fully analogue photonic reservoir computer. *Scientific Reports* 6: 22381.
16. Benedetto, Sergio, and Ezio Biglieri. 1999. *Principles of digital transmission: With wireless applications*. Springer Science & Business Media.

17. Singh, Jaspreet, Sandeep Ponnuru, and Upamanyu Madhow. 2009. Multigig bit communication: The ADC bottleneck. In *IEEE international conference on ultra-wideband, 2009. ICUWB 2009*, 22–27. IEEE

18. Sobel, David Amory, and Robert W. Brodersen. 2009. A 1 Gb/s mixed-signal baseband analog front-end for a 60 GHz wireless receiver. *IEEE Journal of Solid-State Circuits* 44 (4): 1281–1289.

19. Feng, Xiaodong, Guanghui He, and Jun Ma. 2010. A new approach to reduce the resolution requirement of the ADC for high data rate wireless receivers. In *2010 IEEE 10th international conference on signal processing (ICSP)*, 1565–1568. IEEE

20. Yong, Su-Khiong, Pengfei Xia, and Alberto Valdes-Garcia. 2011. *60 GHz technology for Gbps WLAN and WPAN: From theory to practice*. Wiley

21. Hassan, Khursheed, Theodore S Rappaport, and Jeffrey G Andrews. 2010. Analog equalization for low power 60 GHz receivers in realistic multipath channels. In *Global telecommunications conference (GLOBE–COM 2010), 2010 IEEE*, 1–5. IEEE

22. Malone, Jerry, and Mark A Wickert. 2011. Practical volterra equalizers for wideband satellite communications with twta nonlinearities. In *Digital signal processing workshop and IEEE signal processing education workshop (DSP/SPE), 2011 IEEE.*, 48–53. IEEE

23. Bauduin, Marc, Anteo Smerieri, Serge Massar, and François Horlin. 2015. Equalization of the non-linear satellite communication channel with an echo state network. In *vehicular technology conference (VTC Spring), 2015 IEEE 81st*, 1–5. IEEE

24. Mathews, V John, and Lee Junghsi. 1994. Adaptive algorithms for bilinear filtering. In *SPIE's 1994 international symposium on optics, imaging, and instrumentation. international society for optics and photonics*, 317–327.

25. Arfken, George B. 1985. *Mathematical methods for physicists*. Orlando, FL: Academic Press

26. IEEE Standard VHDL Language Reference Manual. 1994. In: *ANSI/IEEE Std 1076-1993*.

27. Pedroni, Volnei A. 2004. *Pedroni*. Circuit Design with VHDL: MIT Press.

28. Virtex-6 FPGA DSP48E1 Slice. 2011. UG369. Xilinx Inc.

29. Wang, Jingjing, Hao Zhang, Tingting Lv, and T Aaron Gulliver. 2012. Capacity of 60 GHz wireless communication systems over fading channels. *Journal of Networks* 7.1: 203–209.

30. Duarte, Melissa, Ashutosh Sabharwal, Vaneet Aggarwal, Rittwik Jana, KK Ramakrishnan, Christopher W Rice, and NK Shankaranarayanan. 2014. Design and characterization of a full-duplex multiantenna system for WiFi networks. In *IEEE transactions on vehicular technology* vol 63.3, 1160–1177.

31. Jaeger, Herbert, and Harald Haas. 2004. Harnessing nonlinearity: Predicting chaotic systems and saving energy in wireless communication. *Science* 304: 78–80.

32. Antonik, Piotr, François Duport, Anteo Smerieri, Michiel Hermans, Marc Haelterman, and Serge Massar. 2015. Online training of an optoelectronic reservoir computer. In *APNNA's 22th international conference on neural information processing*. Vol. 9490, 233–240. LNCS

33. Hermans, Michiel, Joni Dambre, and Peter Bienstman. 2015. Optoelectronic systems trained with backpropagation through time. *IEEE Transactions on Neural Networks and Learning Systems* 26 (7): 1545–1550.

34. Hermans, Michiel, Miguel Soriano, Joni Dambre, Peter Bienstman, and Ingo Fischer. 2015. Photonic delay systems as machine learning implementations. *JMLR* 16: 2081–2097.

35. Antonik, Piotr, Michiel Hermans, François Duport, Marc Haelterman, and Serge Massar. 2016. Towards pattern generation and chaotic series prediction with photonic reservoir computers. In *SPIE's 2016 laser technology and industrial laser conference*. Vol. 9732. 97320B.

# Chapter 3
# Backpropagation with Photonics

This chapter presents an experiment that was not originally planned as part of my thesis. The project was set up when Michiel Hermans joined our team in 2015 with an idea of implementing the backpropagation training algorithm (more on that in Sect. 3.2) in hardware, using our opto-electronic reservoir computer (see Sect. 1.2.4) with one slight modification. After a couple of brainstormings, we decided that the idea would be easier to realise with a FPGA in the setup. I was thus set to assist Michiel Hermans with this experiment, that we successfully completed in March 2016. Michiel Hermans designed the setup and performed most of the measurements, while I built the experiment and programmed the FPGA.

The content of this chapter is based on our paper [1]. The paper itself, being a Letter, is quite succinct, but the Supplementary Material covers most of the theoretical and experimental aspects of this work. The only part not covered in the paper—the FPGA design—is presented in this thesis in Sect. 3.4.

## 3.1 Introduction

One of the main difficulties when using nonlinear dynamical systems, such as neural networks, is to train their internal parameters. The backpropagation (BP) algorithm [2, 3] is one of the most important algorithms in this area, and is behind the remarkable successes achieved in the field of deep learning in the last decade [4]. The simple idea behind the BP algorithm is to compute the derivative (or gradient) of a cost function in the parameter space of the system. The gradient is then subtracted from the parameters themselves in order to reduce the cost function. This process is repeated until the cost function no longer reduces. We will cover BP much more in detail in Sect. 3.2.

In hardware implementations of neural networks the training of internal parameters is also key and the use of the BP algorithm is highly beneficial in order to

© Springer International Publishing AG, part of Springer Nature 2018

P. Antonik, *Application of FPGA to Real-Time Machine Learning*,
Springer Theses, https://doi.org/10.1007/978-3-319-91053-6_3

improve performance [5, 6]. However implementing the BP algorithm in hardware systems can be difficult because of the need of an accurate model to compute the gradient and because of the resources necessary to run the BP algorithm. Remarkably, in certain cases it can be implemented physically on the system it is optimising [7]. The basic idea behind this advance is to use a slightly modified version of the system for propagating error signals backwards, i.e. for running the BP algorithm. Such self-learning computing systems could be highly advantageous, as any gain in terms of processing speed or limited power consumption will also apply to the training phase. Furthermore having the same hardware computing the BP algorithm eliminates, to a large extent, the need for an accurate model of the system. This idea may conceivably also have implications for biological neural networks, as these are physical system that—using mechanisms that are not yet well understood—can both compute and carry out their own training process. A proof of concept experiment in which physical BP was tested on a simple task was reported in [7]. However, it left open the question of whether the algorithm, with all the imperfections inherent in an experiment, can provide the same improvement in performance as numerical approaches [5, 6].

In this experiment we implemented the BP algorithm physically on the slightly modified opto-electronic reservoir computer, introduced in Sect. 1.2.4. The key innovation is to modify the system by adding a photonic component capable of implementing both the nonlinearity and its derivative, so that it can be used both as signal processor and to perform the BP algorithm. We tested our system on several tasks considered hard in the machine learning community, including a real world phoneme recognition task, obtaining state of the art results when the BP algorithm is used. This experiment thus demonstrates the full potential of physical BP. It constitutes an important step towards self-learning hardware, with potential applications towards ultra-fast, low energy consumption, computing systems.

## 3.2 Backpropagation Through Time

Backpropagation through time, or simply backpropagation, is a time-hon-oured method for training recurrent neural networks [2, 3]. Essentially, one defines a cost function based on the desired network output and, using the chain rule, determines the gradient of this cost function with respect to the internal parameters of the system (weights). The term "backpropagation" stems from the fact that, due to the recursion in RNNs, computing the gradient involves propagating an error signal backward through the system updates, i.e., backward in time [5].

In simple words, the algorithm works as follows. The RNN is fed with an input $I$, that propagates through the network, from input to output neurons, in a certain number of timesteps (updates). Using a set of readout weights, the output $O$ is computed. It is then compared with the target output $T$ and an error $E$ is computed. At this point, the network is inverted, figuratively. The error signal $E$ is propagated backwards through the linearised network, from output to input neurons, in the same manner as the initial input signal $I$ was propagated in the forward direction. That is, output neurons act as inputs and input neurons are considered as system outputs. This allows to compute the error on each neuron, i.e. how its value should be changed (to first order in the error) in order to get closer to the desired value $T$. Obviously, the neurons states cannot be "corrected", but the weights can be. These errors are thus used to compute corrections for the input, internal and readout weights, so that that the system output $O$ for a given input $I$ gets closer to the target $T$. The (usually small) corrections are applied iteratively until the error between $O$ ant $T$ can no longer be decreased.

And now we will explicitly derive the equations behind the backprop algorithm. This is probably the most computationally intense section of this thesis. Big thanks to Michiel Hermans for doing all the hard work!

### 3.2.1 General Idea and New Notations

Before we dive into calculations, let me present the notations used in this chapter, matching the original paper [1].

In typical RC tasks, the goal is to map an input sequence $s_i$ (where $i \in \{1, \ldots, L\}$, with $L$ the total sequence length) to an output sequence $y_i$, which has target values $y_i^*$, for example a speech signal to a sequence of labels. In order to use delay-coupled systems as reservoir computers, the discrete time input sequence $s_i$ is encoded into a continuous time function $z(t)$ by the input mask $m(r)$ and bias mask $m_b(r)$, where $r \in [0, T]$, with $T$ the masking period, as follows

$$z(t) = z(iT + r) = m(r)s_i + m_b(r) . \tag{3.1}$$

Our opto-electronic reservoir computer with a sine nonlinearity obeys the equation

$$a(t + D) = \mu \sin (a(t) + z(t)) , \tag{3.2}$$

where $a(t)$ is the state variable and $D$ is the delay. The factor $\mu$ corresponds to the total loop amplification. Equation 3.2 can be seen as a special case of the Ikeda delay differential equation [8].

One then needs to map the continuous time state variable $a(t)$ to a discrete time output sequence $y_i$. This is performed using an output mask $u(r)$ where $r \in [0, T]$ and a bias term $u_b$ as follows:

$$y_i = \int_0^T dr \, a(iT + r)u(r) + u_b . \tag{3.3}$$

In the RC paradigm the input mask is typically chosen randomly, and the output mask $u(r)$ and $u_b$ is determined by solving a linear system of equations which minimises the mean square error $C$ between the desired and actual output $C = \langle (y_i - y_i^*)^2 \rangle_i$.

The goal of applying error backpropagation to RC is to optimise both the input and output masks $m(r)$, $m_b(r)$, $u(r)$ and $u_b$, knowing the reservoir state $a(t)$, and the desired output $y_i^*$. To this end one needs the gradient of the error function $C = \langle (y_i - y_i^*)^2 \rangle_i$ with respect to the masks, given by

$$\bar{e}(iT + r) = e_i u(r) , \tag{3.4}$$

$$e(t - D) = J(t)\,(e(t) + \bar{e}(t)) , \tag{3.5}$$

$$J(t) = \mu \cos\,(a(t) + z(t)) , \tag{3.6}$$

$$\frac{dC}{dm(r)} = \sum_i e(iT + r)s_i , \tag{3.7}$$

$$\frac{dC}{dm_b(r)} = \sum_i e(iT + r) , \tag{3.8}$$

where $\bar{e}(t) = \partial C/\partial a(t)$ is a continuous time signal and, as above, $i \in \{1, \ldots, L\}$ and $r \in [0, T]$. One can then iteratively improve the masks so as to lower $C$. In the following sections we will explicitly derive the above equations.

## 3.2.2  Setting Up the Problem

We wish to find the gradient of a cost function $C$ with respect to the parameters that can be optimised. In order to achieve this we have to use the chain rule through all the dependencies that describe the system. Figure 3.1 gives a schematic of how the forward and backward equations must be implemented experimentally. Figure 3.2 depicts the information flow in the forward and backward systems.

We first recall the relevant equations describing the forward system. The input signal $z(t)$, formed by concatenating the input masks weighted with the current input sample $s_i$ can be rewritten as

$$z(t) = s_{\lceil t/T \rceil} m(t \bmod T) + m_b(t \bmod T), \tag{3.9}$$

where $\lceil . \rceil$ indicates the ceiling function, so that $\lceil t/T \rceil = i$ gives the index of $s_i$ corresponding to the time $t$. We use the modulo operation in the argument of the input masks to indicate that the masks are repeated over time. Next we write down the expression for the reservoir state $a(t)$:

$$a(t + D) = \mu \sin\,(a(t) + z(t)) . \tag{3.10}$$

**Fig. 3.1** **a** Schematic depiction of the forward system as given be Eqs. (3.9, 3.10, 3.11). **b** Schematic depiction of the backward system as given by Eqs. (3.15, 3.20), where $q$ is the backwards time

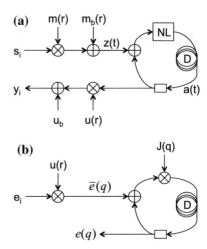

Finally, we can write the formula for the output instances $y_i$ as follows:

$$y_i = u_b + \int_0^T dr\ u(r)a_i(r), \tag{3.11}$$

with $a_i(r) = a(r + (i-1)T)$, the $i$-th segment of the recording of $a(t)$.

In what follows, for the sake of generality and of simplicity of notation, we take the input and output masks to be continuous functions of time. We denote functional derivatives with respect to time dependent functions as ordinary derivatives. The case, relevant to practical implementations, in which the masks depend on a finite number of parameters, is discussed in Sect. 3.3.2. For simplicity in the derivations we will assume, unless indicated otherwise, that all variables, both in continuous time $t$ and discrete time $i$, are defined for $i$ and $t$ going from $-\infty$ to $\infty$. If we have a specific finite input sequence $s_i$ with $i \in \{1, \dots, L\}$, we simply extend this beyond these bounds assuming that all extra $s_i$ are equal to zero. Similarly, we assume that $z(t)$ is zero if $\lceil t/T \rceil \notin \{1, \dots, L\}$. Subsequently, if we sum or integrate over $i$ or $t$ without indicating limits, this indicates a summation or integration from $-\infty$ to $\infty$. In practice it turns out that if we only have a finite sequence, we only need to compute states over its corresponding time span. Similarly, when performing backpropagation, we only need to compute backwards over the same time span. All states outside of this interval do not influence the gradient computation, which means there are no problems in considering only finite intervals. This matters, as in realistic training scenarios we typically train on relatively short sequences (in the case of the present paper of length 100).

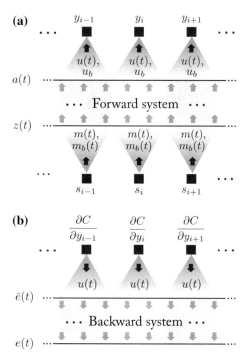

**Fig. 3.2  a** Schematic depiction of information flow when the system is used in the forward direction. On the bottom, the input sequence $s_i$ is converted to a continuous-time signal $z(t)$ (with time running from left to right). Each instance in the sequence is multiplied with the finite-length masking signal $m(t)$ and added to $m_b(t)$. These sequences are then concatenated in time to form $z(t)$, the input to the forward system. The output $a(t)$ of the forward system is then converted into an output sequence $y_i$ by segmenting $a(t)$ in time, and multiplying the segments with the output masks $u(t)$, and integrating over each of them. **b** Schematic depiction of the information flow in the "backwards" mode. The derivatives $\partial C/\partial y_i$ are used as an input sequence. They are multiplied by $u(t)$ which now plays the role of input mask. This yields the signal $\bar{e}(t)$ that serves as input for the backward system. The output of the backward system is $e(t)$

### 3.2.3  Output Mask Gradient

For the output masks we can write

$$\frac{dC}{du(r)} = \sum_i \frac{\partial C}{\partial y_i} \frac{dy_i}{du(r)}. \tag{3.12}$$

For example, if the cost function we wish to minimise is the squared error over the interval of the input sequence

$$C = \sum_{i=1}^{L} (y_i - y_i^*)^2,$$

$$e_i = \frac{\partial C}{\partial y_i} = 2(y_i - y_i^*) \text{ for } i \in \{1, \ldots, L\},$$

$$\frac{\partial C}{\partial y_i} = 0 \text{ for } i \notin \{1, \ldots, L\}.$$

The second factor in Eq. (3.12) we can get from Eq. 3.11:

$$\frac{dy_i}{du(r)} = a_i(r),$$

such that the gradient for the output mask $u(t)$ is simply given by

$$\frac{dC}{du(r)} = \sum_i \frac{\partial C}{\partial y_i} a_i(r),$$

or, given the fact that $\partial C / \partial y_i = 0$ outside the interval in which the sequence is defined

$$\frac{dC}{du(r)} = \sum_{i=1}^{L} \frac{\partial C}{\partial y_i} a_i(r). \qquad (3.13)$$

Similarly we find that

$$\frac{dC}{du_b} = \sum_{i=1}^{L} \frac{\partial C}{\partial y_i}.$$

### 3.2.4  Input Mask Gradient

The case of the input masks is more involved. Working out the chain rule we find

$$\frac{dC}{dm(r)} = \sum_i \frac{\partial C}{\partial y_i} \frac{dy_i}{dm(r)}.$$

$$= \sum_i \frac{\partial C}{\partial y_i} \int dt' \frac{\partial y_i}{\partial a(t')} \frac{da(t')}{dm(r)}.$$

$$= \int dt' \, \bar{e}(t') \frac{da(t')}{dm(r)}, \qquad (3.14)$$

where we have used

$$\bar{e}(t') = \frac{\partial C}{\partial a(t')} = \sum_i \frac{\partial C}{\partial y_i} \frac{\partial y_i}{\partial a(t')} \, .$$

From Eq. 3.11 we can obtain (using a modulo function in the argument of $u(r)$)

$$\frac{\partial y_i}{\partial a(t')} = \delta_{i,\lceil t'/T\rceil} u(t' \bmod T),$$

i.e., equal to zero when $t'$ did not fall in the segment of time used to produce $y_i$, and equal to the output mask otherwise. This yields

$$\bar{e}(t') = u(t' \bmod T) \frac{\partial C}{\partial y_{\lceil t'/T\rceil}}$$

$$= u(r)e_i \tag{3.15}$$

where $r = t' \bmod T$ and $i = \lceil t'/T \rceil$. In other words, $\bar{e}(t)$ is produced by masking the sequence $\partial C/\partial y_i$ with the output mask $u(r)$.

The second factor in Eq. 3.14 we work out as follows. Using the chain rule we get

$$\frac{da(t')}{dm(t)} = \int dt'' \frac{da(t')}{dz(t'')} \frac{dz(t'')}{dm(t)}. \tag{3.16}$$

and

$$\frac{da(t')}{dz(t'')} = \frac{\partial a(t')}{\partial z(t'')} + \int dt''' \frac{\partial a(t')}{\partial a(t''')} \frac{da(t''')}{dz(t'')}.$$

From Eq. 3.10 we obtain the partial derivatives

$$\frac{\partial a(t')}{\partial z(t'')} = \frac{\partial a(t')}{\partial a(t'')}$$

$$= \mu \delta(t' - t'' - D) \cos(a(t' - D) + z(t' - D)),$$

Or, more compactly,

$$\frac{\partial a(t')}{\partial z(t'')} = \delta(t' - t'' - D) J(t'),$$

with

$$J(t') = \mu \cos(a(t' - D) + z(t' - D)).$$

This yields

$$\frac{da(t')}{dz(t'')} = J(t') \left[ \delta(t' - t'' - D) + \frac{da(t' - D)}{dz(t'')} \right].$$

By filling in the expression for $da(t' - D)/dz(t'')$ recursively we can write this as

$$\frac{da(t')}{dz(t'')} = \sum_{i=0}^{\infty} \left[ \delta(t' - t'' - iD) \prod_{j=0}^{i-1} J(t' - jD) \right]. \tag{3.17}$$

By filling in Eq. 3.17 in Eq. 3.16, and inserting the result in Eq. 3.14 we obtain

$$\frac{dC}{dm(r)} = \int dt' \, dt'' \, \bar{e}(t') \sum_{i=0}^{\infty} \delta(t' - t'' - iD) \prod_{j=0}^{i-1} J(t' - jD) \frac{dz(t'')}{dm(r)}. \tag{3.18}$$

We can solve the integral over $t'$ explicitly. We denote this by $e(t'')$:

$$e(t'') = \int dt' \, \bar{e}(t') \sum_{i=0}^{\infty} \delta(t' - t'' - iD) \prod_{j=0}^{i-1} J(t' - jD)$$

$$= \sum_{i=0}^{\infty} \bar{e}(t'' + iD) \prod_{j=0}^{i-1} J(t'' + (i - j)D)$$

$$= \sum_{i=0}^{\infty} \bar{e}(t'' + iD) \prod_{j=1}^{i} J(t'' + jD). \tag{3.19}$$

It is straightforward to prove that $e(t)$ is equal to Eq. 3.5 (with arguments shifted by $D$)

$$e(t) = J(t + D)(e(t + D) + \bar{e}(t + D)). \tag{3.20}$$

Indeed, if we recursively fill in the expression for $e(t + D)$ in Eq. 3.20, we obtain Eq. 3.19. Using this we can reduce Eq. 3.18 to

$$\frac{dC}{dm(r)} = \int dt'' \, e(t'') \frac{dz(t'')}{dm(r)}.$$

From the expression of $z(t)$ we find that

$$\frac{dz(t'')}{dm(r)} = \delta(t'' \bmod T - r) s_{\lceil t/T \rceil}.$$

Inserting this we can find the final expression for the gradient for the input mask

$$\frac{dC}{dm(r)} = \sum_i s_i e_i(r),$$

or, again using the fact that we defined $s_i = 0$ for $i \notin \{1, \ldots, L\}$,

$$\frac{dC}{dm(r)} = \sum_{i=1}^{L} s_i e_i(r), \tag{3.21}$$

with $e_i(r) = e(r - (i-1)T)$, the $i$-th segment of the time trace of $e(t)$. Similarly for $m_0(t)$ we can write

$$\frac{dC}{dm_b(r)} = \sum_{i=1}^{L} e_i(r). \tag{3.22}$$

### 3.2.5  Multiple Inputs/Outputs

The above explanation is easily extended to multiple input and output dimensions. Suppose that we have a multivariate time series $\mathbf{s}_i$, where the $k$-th element at time step $i$ is denoted by $\mathbf{s}_i[k]$. We can then easily construct $z(t)$ by defining as many input masks $m_k(t)$ as there are input dimensions and adding them all up

$$z(t) = \sum_k \mathbf{s}_{\lceil t/T \rceil}[k] m_k(t \bmod T) + m_b(t \bmod T),$$

The desired output can similarly exist of a multivariate time series with elements $\mathbf{y}_i^*[l]$. To produce an output $\mathbf{y}_i[l]$ we simply define an output mask $u_l(t)$ and bias $u_l^0$ for each output channel

$$\mathbf{y}_i[l] = u_l^0 + \int_0^T dt\, u_l(r) a_i(r).$$

The same procedure can now be used to determine the gradients with respect to the multivariate input and output masks. We find

$$\frac{dC}{du_l(r)} = \sum_{i=1}^{L} \frac{dC}{d\mathbf{y}_i[l]} a_i(r), \tag{3.23}$$

and

$$\frac{dC}{du_l^0} = \sum_{i=1}^{L} \frac{dC}{d\mathbf{y}_i[l]}. \tag{3.24}$$

The source of the BP equation is now

$$\bar{e}(t') = \sum_l u_l(t' \bmod T) \frac{dC}{d\mathbf{y}_{\lceil t'/T \rceil}[l]}, \tag{3.25}$$

the recurrence for the error $e(t)$, Eq. (3.20), is unchanged, and one has

$$\frac{dC}{dm_k(r)} = \sum_{i=1}^{L} s_i[k]e_i(r).$$

## 3.3  Experimental Setup

In order to use the same hardware for both the signal processing and its own training, one exploits the very close analogy between Eqs. 3.1 and 3.4—both are formed in the same way from a discrete time sequence, multiplied by a periodic mask—as well as the very close analogy between Eqs. 3.2 and 3.5—both are delay systems. However the equation for $e(t)$ depends on future values, so it needs to be solved backwards in time. In practice one time-inverts $\bar{e}(t)$ and $J(t)$ before computing $e(t)$ to obtain a linear delayed equation

$$e(q + D) = J(q)\,(e(q) + \bar{e}(q)),\tag{3.26}$$

where we use $q$ instead of $t$ to remind oneself that we are dealing with time-inverted signals. We also note that $J(t)$, the derivative of the nonlinear function, is a cosine, which can also be implemented using the intensity modulator. Although this property of the sine function is key for this experiment, other types of nonlinearity can be implemented in analogue hardware (see Sect. 3.7).

In this work we have shown how Eqs. 3.2 and 3.26 can be realised using the same physical setup, depicted in Fig. 3.3. It looks very similar to the opto-electronic reservoir computer discussed in Sect. 1.2.4. The key difference (and innovation) is the use of two dual input/dual output Mach-Zehnder modulators (MZMs), which allows to implement both Eqs. 3.2 and 3.26 using the same physical system. Taking into

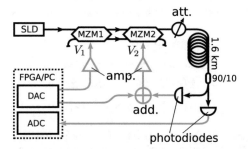

**Fig. 3.3** Schematic representation of the experimental system. SLD: superluminescent diode; MZM1 and MZM2: dual input/dual output Mach-Zehnder Modulators; $V_1$ and $V_2$: driving voltages of the MZMs; att.: programmable optical attenuator; add.: electrical combiner; amp.: pulse amplifier

account the incoherence of light in the two branches between the modulators, the
output of the upper branch of MZM2 (see Fig. 3.3) can be found to be

$$I_2^+ = \frac{I_0}{2} \left[1 + \sin(V_1/V_0)\sin(V_2/V_0)\right], \qquad (3.27)$$

where $I_0$ is the input intensity in the upper branch of MZM1, $V_1$ and $V_2$ are the
driving voltages and $V_0$ a constant depending on the MZM. The computational details
will be presented below, in Sect. 3.3.1. In the forward mode, we choose $V_1/V_0 =
\pi/2$. The transfer function thus acts as a sinusoidal function for the input argument
$V_2/V_0 = a(t) + z(t)$. The constant offset $I_0/2$ is removed by the high-pass filter
of the amplifier, that drives the MZM. Therefore, once the loop is closed, we end
up with Eq. 3.2. In the backward mode we drive MZM1 with a voltage $V_1/V_0 =
a(q) + z(q) + \pi/2$, and MZM2 with a signal proportional to $\bar{e}(q) + e(q)$, but scaled
down sufficiently such that $\sin(V_2/V_0) \approx V_2/V_0 = \bar{e}(q) + e(q)$, which gives the
desired functionality for the adjoint system Eq. 3.26.

In order to train our reservoir computer, we first choose a value of $\mu$ close to
the threshold for instability. We then iterate the following three steps for (typically)
several thousands of iterations, during which performance slowly improves until it
converges:

(1) We take the training data (typically a small subsequence of the complete set), and
    convert it to $z(t)$ using the input masks. We feed this signal to the experimental
    setup, physically implementing Eq. 3.2. Next, we measure and record the signal
    $a(t)$, and generate an output sequence $y_i$ using the output masks.
(2) From the output and the desired target values we compute the sequence $e_i =
    \partial C/\partial y_i$ at the output, and convert it to $\bar{e}(t)$, now using the output mask as
    an input mask. Next we time-invert it and feed it back into the experimental
    setup. Simultaneously we drive the first MZM with the (time-inverted) signal
    $a(q) + z(q)$ in order to implement the online multiplication with $J(q)$. We record
    the response signal $e(q)$.
(3) From the recorded signals $a(t)$ and $e(t)$ we obtain the gradients for the masking
    signals, which we use to update the input and output masks,

$$m(r) \leftarrow m(r) - \eta\, dC/dm(r),$$
$$m_b(r) \leftarrow m_b(r) - \eta\, dC/dm_b(r),$$
$$u(r) \leftarrow u(r) - \eta\, dC/du(r),$$
$$u_b \leftarrow u_b - \eta\, dC/du_b, \qquad (3.28)$$

where $\eta$ is a (typically small) learning rate. In order to speed up convergence we
applied a slightly more advanced variant of these update rules known as Nesterov
momentum [9, 10] (see Sect. 3.5.4 for more details).

The FPGA, depicted in Fig. 3.3, simultaneously generates the voltage signal that
represents $z(t)$ and records the voltage signal representing $a(t)$. It also performs a

minimal signal processing step by selecting and averaging over the middle samples of each masking step (see Sect. 3.3.2 for more details). The remaining processing steps are carried out on a PC. The FPGA operation is discussed further in Sect. 3.4.

Sending and receiving data to and from the FPGA was the main speed bottleneck of the experiment (more on that in Sec. 3.6). Even though a single training iteration lasts only about 0.6 s for the NARMA10 and VARDEL5 task, most of this time is spent on the communication overhead with the PC (buffering). If the entire experiment were to be performed on the FPGA (which is feasible), a single training iteration would take of the order of milliseconds.

### 3.3.1 Online Multiplication Using Cascaded MZMs

We used two back to back dual input/dual output Mach-Zehnder modulators for implementation of both Eqs. 3.2 and 3.26 using the same setup. The main fact we rely on is that the spectrum of the SLD is narrow enough to allow for a large extinction ratio by the MZMs, but is broad enough that the light in the two branches entering MZM2 from MZM1 can be considered incoherent. In the present experiment, the coherence length of the light from the SLD is of the order of a few hundreds of micrometers, which means that a very small difference in path length for the connections in between MZM1 and MZM2 is sufficient to make the two signals incoherent.

Consider the operation of a single MZM, schematised in Fig. 3.4. The intensities of the incoming light sources are denoted by $I_a$ and $I_b$, and the MZM is driven by a voltage $V$, which is the sum of a constant bias voltage $V_B$ and a fast voltage signal $V_{RF}$. The bias voltage was omitted in the main text to avoid confusion. Taking into account the incoherence between the two input signals, the intensities of the output branches ($I^+$ and $I^-$) are given by

$$I^+ = I_a \frac{1 + \sin(V/V_0)}{2} + I_b \frac{1 - \sin(V/V_0)}{2},$$
$$I^- = I_a \frac{1 - \sin(V/V_0)}{2} + I_b \frac{1 + \sin(V/V_0)}{2}, \qquad (3.29)$$

with $V_0$ a constant depending on the MZM.

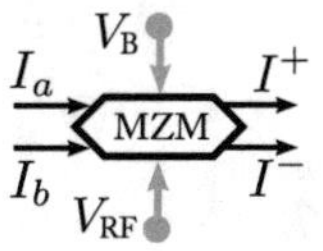

**Fig. 3.4** Schematic representation of a dual input/dual output Mach Zehnder modulator. The MZM is driven by the sum of two input voltages: one constant bias voltage $V_B$ and a fast signal $V_{RF}$

It is now easy to model the output of the two cascaded MZMs. Suppose the source has an intensity $I_0$, and no light enters the second input of MZM1. And suppose MZM1 and MZM2 receive voltages $V_1$ and $V_2$, respectively. The output intensities $I_1^+$ and $I_1^-$ of MZM1 are given by

$$I_1^+ = I_0 \frac{1 + \sin(V_1/V_0)}{2},$$

$$I_1^- = I_0 \frac{1 - \sin(V_1/V_0)}{2}. \tag{3.30}$$

The intensity $I_2^+$ at the first output branch of MZM2 is then

$$I_2^+ = I_0 \frac{(1 + \sin(V_1/V_0))(1 + \sin(V_2/V_0))}{4} \tag{3.31}$$

$$+ I_0 \frac{(1 - \sin(V_1/V_0))(1 - \sin(V_2/V_0))}{4} \tag{3.32}$$

$$= \frac{I_0}{2} [1 + \sin(V_1/V_0) \sin(V_2/V_0)]. \tag{3.33}$$

In the experiment MZM1 receives a constant bias signal on top of an RF driving signal, such that $V_1/V_0 = \pi/2 + V_1'/V_0$, with $V_1'$ the RF signal. We can thus write:

$$I_2^+ = \frac{I_0}{2} [1 + \cos(V_1'/V_0) \sin(V_2/V_0)].$$

We use the setup in two modes. In the forward mode, $V_1' = 0$, so that the cascaded MZMs behave as

$$I_2^+ = \frac{I_0}{2} [1 + \sin(V_2/V_0)],$$

i.e., the transfer function acts as a sinusoidal function for the input argument $V_2/V_0$, which is equal to the sum of the input signal $z(t)$ and the system state $a(t)$. Note that a constant offset $I_0/2$ is added to the output. We use, however, amplifiers with a high-pass filter to drive the MZMs, which remove the DC offset. Therefore, once the loop is closed, this constant bias is removed, and we effectively end up with Eq. 3.2.

In the backwards mode, we drive MZM1 with a voltage $V_1'$ proportional to $a(q - D) + z(q - D)$. MZM2 is driven with a signal proportional to $\bar{e}(q) + e(q)$, but scaled down sufficiently such that $\sin(V_2/V_0) \approx V_2/V_0 = \bar{e}(q) + e(q)$. This means that in the backwards mode we can write

$$I_2^+ = \frac{I_0}{2} [1 + \cos(a(q + D) + z(q + D)) (\bar{e}(q) + e(q)],$$

which is (up to the constant bias, and the factor $\mu$ which is imposed later by the optical attenuator) the desired functionality for the adjoint system (given by Eq. 3.26).

## 3.3.2 Mask Parametrisation

While the BP theory, discussed in Sect. 3.2, is generally valid for continuous-time signals, an experimental setup is limited by the finite bandwidth of the DAC/ADC, and the analogue electronic parts. To make sure that these effects play a limited role, we parametrise the input and output masks as piecewise constant functions, which has been common practice for reservoirs of this type [11]. To this end we divide the delay $D$ into an integer number $N_D$ of equal time segments, called *masking steps*. Next, we ensure that the masking period $T$ has a total duration that also contains an integer number $N_T$ of masking steps, in our case one less than the delay: $N_T = N_D - 1$. This allows for the mixing of the states over time, as detailed in [11].

The input and output masks are picked to be constant for the duration of each masking step. This implies that $z(t)$ is piecewise constant. The fact that both $T$ and $D$ are an integer number of masking steps makes that changes in $a(t)$ only occur in between the masking steps, i.e., they are synchronised with the masking steps, and this is valid for the backwards pass too. In short, $a(t)$, $\bar{e}(t)$ and $e(t)$ are all piecewise constant signals, with values that remain constant during each masking step.

In practice this allows us to reduce effects of noise by averaging the signals representing $a(t)$ and $e(t)$ over several measuring samples during a single masking step. Typically we pick a set of samples from the middle of each masking step, and discard those at the beginning and the end as they may contain artefacts caused by the limited bandwidth of the ADC. More importantly, it allows us to make a discrete time approximation of the entire system. For example, let us consider Eq. 3.11. The mask $u(t)$ is made up of $N_T$ constant segments of equal length, with values during the segments denoted $u_k$. Similarly, each segment $a_i(r) = a(t - (i-1)T)$ is piecewise constant, with values we can for example denote with $a_k^i$. The integral reduces to

$$y_i = u_b + \sum_{k=1}^{N_T} a_k^i u_k,$$

(where we absorbed the factor $T$ that emerges from the integration into the values $u_k$). Each particular value $a_k^i$ can be interpreted as the state of the $k$-th "neuron" or "node" state during the $i$-th instance of the input sequence. We can still use the expressions for the gradients in Eqs. 3.13, 3.21 and 3.22. Indeed, by construction, the gradient for the output mask $u(t)$ for the duration of a single masking step is a constant (as $a(t)$ remains constant over the segment). The same holds for the gradients for the input masks. This implies that $u(t)$ and $m(t)$ remain piecewise constant during training, and we can in practice describe them simply as lists of values instead of a continuous-time functions.

Note that the choice of dealing with bandwidth limitations by using piecewise constant functions is not the only possible avenue. One alternative would be to impose bandwidth constraints on $m(t)$ and $u(t)$, such that the finite signal generator bandwidth and sampling rates form no obstacle in treating the setup as a continuous-time setup. We chose the piecewise-constant constraint as it is more directly related to existing implementations of delay-coupled electro-optical signal processors, and it allows to identify a specific number of "virtual nodes" (the number of segments within the masking period $T$). In other words, the choice of $N_T$ determines the "complexity", or the number of degrees of freedom of the system.

## 3.4 FPGA Design

The FPGA design is my key contribution to this project. As will be shown below, it allowed to significantly speed up the experiment, which is crucial for iterative algorithms. The design itself is much simpler than the previous one (see Sect. 2.5): it performs no computations and only basic signal processing on the acquired data. The main novelty is the Ethernet interface with the computer, that took me some time to develop (more on that in Sect. 3.6). It alows a much faster data transfer than the archaic UART, used previously.

The simplified schematics of the design is depicted in Fig. 3.5. Rectangular boxes represent modules (entities), and rounded rectangles stand for electronic components: the FMC151 daughtercard, the Marvell Alaska PHY device (88E1111) for Ethernet communications (ETH), the Hewlett Packard 8648A clock generator, the PC, running Matlab, and the opto-electronic experiment.

The operation of the FPGA is controlled from Matlab through a Gbit Ethernet connection. Data and various commands, such as memory read/write or state change, are encapsulated into standard UDP packages. The `Ethernet` module interfaces the Marvell Ethernet PHY chip with the rest of the design, receives the UDP packets (frames) and decodes the commands and the data. It also ensures data transmission from the FPGA to the computer.

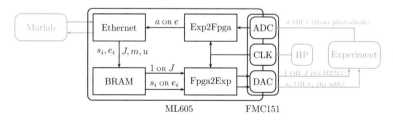

**Fig. 3.5** Simplified schematics of the FPGA design. Modules (entities) are represented by rectangular boxes, electronic components are shown with rounded rectangles. External hardware, such as the computer, running Matlab, the opto-electronic reservoir and the external clock generator are shown in grey. During the forward pass, the FPGA sends the inputs $s_i$ to the electrical combiner (labelled "add." in Fig. 3.3), drives the MZM1 with a constant signal equal to 1 to keep it transparent, and records the reservoir states $a$ from the readout photodiode. During the backward pass, the MZM1 is driven by the Jacobian $J$, the combiner is fed with the errors $e_i$ and the error signal $e$ is recorded

The UDP packets are read at the PHY level on the FPGA (see [12] for a reference on the OSI model). For simplicity, the board does not have a MAC nor an IP address. Matlab sends the packets to a random address, since the cable only connects two devices. The design discards the PHY, TCP and UDP headers from the received packet and only reads the data. It also ignores the checksum at the end of the frame. The first bytes are interpreted as commands and the following, if any, are considered as data and written in the FPGA memory.

Blocks of Random-Access Memory (BRAM) are used to store data, such as the inputs $s_i$ and $e_i$ for the forward and backward runs, respectively, the Jacobian $J$, as well as the input and output masks $u$ and $m$.

Sending packets to the computer is a more challenging process, as the addresses (MAC, IP and port number) must match the computer settings for the data to reach Matlab. Moreover, a 32-bit Cyclic Redundancy Check (also known as the checksum) must be appended to the frame for the computer to accept the packet (and not drop it as a corrupt frame). I assigned an arbitrary IP to the computer and used the default UDP port in Matlab. These addresses are hard-coded in the FPGA design, and the CRC is computed inside the Ethernet module.

The FMC151 daughtercard outputs two 14-bit signals from the ADCs and receives two 16-bit signals for the DACs. It is also used to deliver a clock signal from an external clock generator, that produces a high-precision signal, allowing to synchronise the FPGA with the delay loop of the experimental setup. This clock signal was generated by the Hewlett Packard 8648A signal generator.

The Fpga2Exp module controls the two signals sent to the opto-electronic reservoir through the dual-channel DAC. During the forward pass, it generates the masked input signal $m \times s_i$ by multiplying the inputs $s_i$ by the mask $m$, both being read from the BRAM. During the backward pass, it outputs $u \times e_i$ through one channel, and $J$ through the other.

The reservoir states $a(t)$, as well as the error signal $e(q)$ from the experiment are sampled and averaged by the Exp2Fpga module. The number of samples depends on the task and the reservoir size, as will be discussed in Sect. 3.5. The data is buffered in blocks of RAM (not shown here) and then encapsulated into UDP packages in the Ethernet module and transfered to the computer throught Ethernet.

Since the FPGA performs minimal calculations in this experiment, a legitimate question comes to mind: do we need a FPGA at all? And the answer is yes, we do. Since BP is an iterative process, the experiment needs to be repeated a large number of times for the training to be completed. As will be shown in Sect. 3.5, this number varies from $10^4$ for NARMA and VARDEL up to $10^6$ for TIMIT tasks. Without the FPGA, as in [11], the opto-electronic experiment can be run in approximately 30 s. That is, one training cycle for the NARMA task would have taken about 4 days. With the FPGA, one run takes 0.6 s, cutting the duration of a training cycle to a more manageable time of 2 h.

## 3.5   Results

We validated the experimental setup on three time series processing task.

### 3.5.1   Tasks

We consider first of all the NARMA10 task, introduced in Sect. 1.1.4.2. The second task we will call VARDEL5 (from variable delay). Here, the input sequence consists of independent and identically distributed digits drawn from the set $\{1, 2, 3, 4, 5\}$. The desired output is then given by $y_i^* = s_{i-s_i}$, i.e., the goal is to retrieve the input instance delayed with the number of time steps given by the current input. As a performance metric for NARMA10 and VARDEL5 we use the normalised root mean square error (NRMSE), which is given by

$$\text{NRMSE} = \sqrt{\frac{\langle (y_i - y_i^*)^2 \rangle_i}{\langle (y_i^*)^2 \rangle_i}}.$$

The NRMSE varies between 0 (perfect match), and 1 (no relation between output and target).

The third task is a frame-wise phoneme labelling task. We use the TIMIT dataset [13], a speech dataset in which each time step has been labelled with one of 39 phonemes. The input data is high-dimensional (consisting of 39 frequency channels), and the desired output is one of (coincidentally) 39 possible output classes. The goal is to label each frame in a separate test set. Consequently, the performance metric is now the classification error rate, i.e., the fraction of misclassified phonemes in the test set. Note that the masking scheme and BP algorithm is easily extended to multidimensional in—and output sequences (see Sect. 3.2.5). The TIMIT task has been studied before in the context of RC, which has shown it to be challenging, typically requiring extremely large reservoirs to obtain competitive performance [14, 15].

For all these tasks we compared performance of the fully trained system to traditional RC, where we kept the input and bias masks fixed and random, and only optimised their global scaling and the feedback strength parameter $\mu$. The results are shown in Fig. 3.6. The experimental setup is successful in performing both useful computations, and implementing its own training process. The fully trained system consistently outperforms the RC approach in all tasks considered.

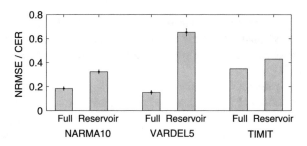

**Fig. 3.6** Comparison of performances for the three tasks under consideration. We show either NRMSE (for NARMA10 and VARDEL5) or the classification error rate (CER) for TIMIT. For each task we show performance for a fully trained systems (Full) versus those trained using the RC paradigm (Reservoir). Error bars indicate standard deviations if available

### 3.5.2 NARMA10 and VARDEL5

In the case of NARMA10 and VAR-DEL5 we divided $T$ into 80 equal time intervals ($N_T = 80$), which allowed us to take 16 samples during each masking step, where we averaged over the middle 8 in order to get piecewise-constant values for $a(t)$ and $e(t)$. We chose the number of training iterations at 10000 and 20000 for VARDEL5 and NARMA10, respectively, chosen heuristically as a trade-off between the time required for an experiment and the final performance (a single iteration lasted approximately 0.6 s).

The cost functions used for NARMA10 and VARDEL5 are the aforementioned sums of squared errors. We repeated the training cycles 10 times, each time with different random input mask initialisations. Output masks were always initialised at zero. For all backpropagation experiments we set the feedback parameter strength parameter $\mu$ effectively equal to one (such that the system is at the "edge of stability"), which we found to give the best performance.

For the NARMA10 task we improve over all previous experimental results. The previous best was published in [16], which reported an NRMSE of 0.249 for 50 virtual nodes, and 0.22 for 300 virtual nodes, whereas here we obtain a NRMSE of 0.185 for 80 nodes (note that in [16] the authors report normalised mean square error (NMSE, presented here in Sect. 1.1.4.2), which is the square of the NRMSE). That result was obtained on an experimental setup that was specially designed to produce a minimal amount of noise (using a passive cavity as a reservoir). The lowest reported experimental NRMSE on a setup equivalent to ours was 0.41 [11]. Note that we obtain a better average performance for the RC setup (NRMSE = 0.32), which is most likely due to the higher number of virtual nodes (80 as opposed to 50 in [11]).

For the VARDEL5 task, we cannot directly compare to literature, however as pointed out in Chap. 5 of Michiel Hermans' thesis [17], this task is an important example of a task that is so nonlinear that it is nearly impossible to solve it with RC. This is confirmed here: the NRMSE of RC is 0.66, indicating that the reservoir has only captured the task on a very rudimentary level. The fully trained system shows a drastically better performance (NRMSE = 0.15). This shows that training the input masks not just allows for better performance on existing tasks, but also allows to tackle tasks that are so intricate that they are considered beyond the reach of traditional RC.

In Fig. 3.7 we depict the masks $m(r)$ and $m_b(r)$ for the RC implementation (when they are chosen at random), and after optimisation using the BP algorithm, for the VARDEL5 task. One sees that the BP algorithm dramatically changes the input masks. In particular, the mask $m$ is very large at some specific values of $r$, and almost zero for other values. This suggests that in some sense what the optimised reservoir is doing is storing the value of the input on specific neurons, and then keeping it in memory for some time, before mixing it nonlinearly with the input several time steps in the future. In Fig. 3.8 we depict how the NRMSE converges over time for the VARDEL5 task, as the BP algorithm slowly improves the input and output masks.

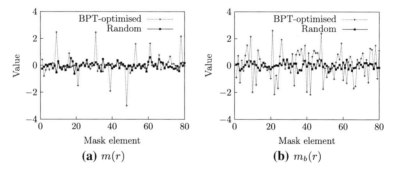

**Fig. 3.7** Comparison of BP-optimised input masks (dotted curves) and random RC masks (solid curves) for the VARDEL5 task, for $m(r)$ (left panel) and $m_b(r)$ (right panel)

**Fig. 3.8** Evolution of the NRMSE during the training process on the VARDEL task. The error falls sharply during the first 300 iterations, and then decreases slowly towards 0.15

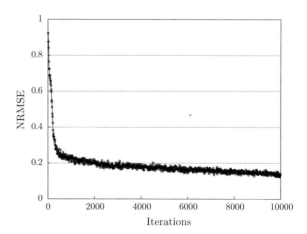

For the reservoir computing results we measured average performance as a function of three scaling parameters: the feedback strength parameter $\mu$ and the scaling of the input mask and bias mask. Once optimal parameters were determined we ensured that the output masks were trained on an unlimited amount of input training data (in practice we observed the test error for increasing amounts of training data, and stopped as soon as the performance no longer improved). This was to ensure that we have a fair comparison to the backpropagation setup, where we generate unlimited amounts of data too. Each experiment was is repeated 10 times, giving rise to the error bars in Fig. 3.6.

### 3.5.3 TIMIT

For the TIMIT task we used $10^6$ training iterations, that took two weeks to complete. For this reason, we only performed a single full training cycle. We picked $\mu$ at a value slightly under one, but we found in simulations that performance did not strongly depend on it for a broad range of values.

Measurement noise plays a smaller role in a classification task such as this one, and we divided $T$ into 200 masking steps, taking 8 samples in each and averaging over the middle 4, thereby increasing the number of virtual nodes $N_T$ while taking into account the hardware constraints (sample rates of the DAC and ADC). Most likely this number can be increased further, for example, using only 4 samples per masking steps and averaging over the middle 2. In practice we are also limited by the relatively slow communication between the PC and the FPGA, and increasing $N_T$ increases the amount of data that needs to be transferred, slowing down the experiment considerably.

The goal here is to minimise a classification error rate, which is not directly differentiable. One possible strategy is simply to try and minimise the MSE between the output and the target labels (1 for the correct class, zero for all others). Classification would then be performed by the winner-take-all approach, where we simply select the output channel with the highest output as the "winner". In practice, using MSE for classification suffers from some drawbacks. Most importantly MSE will put a lot of emphasis on producing the exact target values (close to zero or one), while we are only interested in performance after selecting the highest output. A better approach is to use a so-called softmax function at the output, which converts the output values into a set of probabilities, and minimise the cross-entropy with the target probabilities (again, 1 for the correct class and zero for all others). Details on this strategy can be found in e.g. [18]. In practice, the conversion of the output $\mathbf{y}_i[k]$ into probabilities is performed using the softmax function

$$\mathbf{p}_i[k] = \frac{\exp(\mathbf{y}_i[k])}{\sum_l \exp(\mathbf{y}_i[l])}.$$

The cost function is the cross-entropy

$$C = -\sum_{i=1}^{L}\sum_{k} \mathbf{t}_i[k] \ln \mathbf{p}_i[k],$$

where we denote the target outputs as $\mathbf{t}_i[k]$. It can then be shown that

$$\frac{dC}{d\mathbf{y}_i[k]} = \mathbf{p}_i[k] - \mathbf{t}_i[k],$$

(and again zero if $i \notin \{1, \ldots, L\}$). This means that the error we have at the output takes on virtually the same form as before, only this time there is the intermediary step of the softmax function. Gradients for the output masks are almost the same as before, except for Eqs. 3.23 and 3.24 where we use $\mathbf{p}_i[k] - \mathbf{t}_i[k]$ instead of $\mathbf{y}_i[k] - \mathbf{y}_i^*[k]$. As far as the rest of the BP algorithm goes, we now simply have to mask these "output errors" to produce $\bar{e}(q)$, and the rest plays out exactly as before.

For the RC approach, optimising the parameters (input scaling, bias scaling and feedback gain) on the hardware would be too costly in terms of time. Therefore we optimised them on a PC using a simulation of the physical setup. Once we decided on the parameters, we ran all the TIMIT data through the physical setup and recorded all the responses. Next we trained output weights, again using gradient descent with the above cross-entropy loss.

We obtained a classification error rate of 34.8% for fully trained systems, versus 42.9% for the standard RC approach. These results are only slightly worse than similar experimental results presented in [6], (33.2% for fully trained systems and 40.5% for the RC approach) where 600 virtual nodes were used as opposed as 200 in our case.

### 3.5.4  Gradient Descent

We used stochastic gradient descent to train the masks. For each iteration, we drew a 100 time step sequence to determine a gradient. This sequence was either generated on the fly (in the case of VARDEL5 and NARMA10), or drawn randomly from a training set (TIMIT). Note that as the BP equation is linear, we are in principle free to rescale $\bar{e}$ as we wish. In practice, in order to keep MZM2 in the linear regime, we scaled the input error signal $\bar{e}(t)$ by dividing it by its standard deviation and multiplying with a factor 0.1. The learning rate $\eta$ we choose equal to 0.25 at the start of the training process, after which it drops linearly to zero throughout the course of the experiment. On top of that we use Nesterov momentum with a momentum factor 0.9 to speed up convergence [9, 10]. Nesterov momentum is a heuristic method that finds widespread use in speeding up convergence of stochastic gradient descent. The idea of momentum in gradient descent is to give parameter updates a certain inertia, meaning that previous parameter updates still count in the current one, which helps with overcoming local minima and speeds up convergence. Nesterov momentum is a simple variation of this principle, where the algorithm measures the gradient one update step ahead in order to change its momentum "ahead of time".

### 3.5.5 Robustness

Our work shows that physical BP is robust against imperfections of the physical setup, as illustrated by the following imperfections we were confronted with.

The first imperfection was the high-pass filtering operation of the amplifiers used to drive the MZMs, with a cut-off frequency of 20 kHz. While the high-pass filter is a desirable property (to get rid of voltage bias), this corresponds to a typical time scale of about 8 μs, which is about the same as the loop delay and therefore not negligible. The current experimental setup does not take this filtering operation into account explicitly.

The second imperfection was an imbalance in losses between the two fibres connecting MZM1 with MZM2.

The third imperfection was that the system was not perfectly linear during the backwards pass, since MZM2 is never a perfectly linear system. There is also an important trade-off here. One can reduce the residual nonlinearity by reducing the amplitude of the incoming voltage signal that represents $\bar{e}(t)$. But in turn this also reduces the signal-to-noise ratio of the measurement during the backpropagation phase, such that one needs to find a good balance between these two effects.

All these effects are imperfections inherent to the physically implemented backpropagation phase, but both in simulation and in the actual experiments we found that they only had a very minor impact on the training process and the overall performance.

One parameter that turned out to be crucial was the bias voltage of MZM2. The reason is that even a small offset from an effectively zero level introduces a systematic error in the backpropagation process, such that the measured signal (denoted as $e_c(t)$ to indicate that it is corrupted) becomes

$$e_c(t - D) = J(t)(e_c(t) + \bar{e}(t) + \tilde{e}),$$

with $\tilde{e}$ a constant offset caused by an incorrectly set voltage bias of MZM2. It turned out that, in the experiments, keeping this bias level effectively equal to zero was difficult: very slight drifts on the effective working point of the MZM occurred over the course of minutes/hours. Luckily, the backpropagation is a linear process. This means that we can recover $e(t)$ by performing a second measurement right after measuring $e_c(t)$

$$e_r(t - D) = J(t)(e_r(t) + \tilde{e}),$$

and

$$e(t) = e_c(t) - e_r(t).$$

In other words we simply need to perform two measurements after each other, where in the second one we send a "zero" input error, and subtract this from the first measurement in order to remove the influence of the offset of MZM2. This turned out to solve the problem.

## 3.6   Challenges and Solutions

We had to face two big challenges while working on this project. The first one is FPGA-related and it had an immense impact on the speed of the experiment. The first design I created used the archaic UART interface to transfer data to and from the computer, as in my first experiment (see Chap. 2), with a maximum transfer rate of 128 kbit/s. In this case, one run of the experiment lasted 16 s, with more than 99% of the time spent on data transfer.

Obviously, a faster connection was required. Two options were available at this stage: Gbit Ethernet or PCI Express, both widely used in modern computers. Since the latter is more complex to implement on hardware level, I picked the Ethernet. Although I did manage to make it work in a reasonable time, this was a poor choice on the long run. Ethernet, by its definition, is not suitable for connections between a PC and its periferals. In short, Ethernet is an intrinsically unreliable protocol—the hardware is designed to drop data (sent by chunks called packets or frames) when it is not capable of processing it in time. Therefore, upper-layer protocols, such as TCP/IP, incorporate acknowledgement of received data and requests to retransmit packets that did not make it to the destination. These protocols are usually implemented in software, and designing them in hardware would have taken a lot of time. So I simply skipped them, and hoped for the best.

A good proof of Ethernet's unreliability was found during the first full-length experiments. We discovered that, sometimes, the data received from the FPGA was wrong. The error, spotted after a couple of long and desperate nights, was hard to believe in at first: some packets did not arrive in the same order to the computer as they were sent, meaning that the flow of time-multiplexed data, transmitted by the FPGA, was broken. And this is actually normal for an Ethernet card to switch the order of packets, whenever it chooses to, for…reasons.

A few problems were encountered on the way to a well-working two-way Ethernet connection. In principle, Xilinx offers ready-to-use plug-and-play blocks, called IP cores, that can handle various features of the FPGA. One such core provides the Ethernet connectivity. Sadly, it simply did not work. The source files were missing important pin connections, including the clock signal to the Ethernet PHY chip. Fortunately, after hours of Googling, I found the solution in Satnam Singh's MSDN Blog [19]. Still, the Xilinx IP was too complex for my needs, so I decided to rip off all that was not necessary and only leave the basic interface with Ethernet PHY chip.

Sending data from the computer to the FPGA was easy, as the latter was pro-grammed to disregard addresses and port numbers, and only read the data. That is, the computer could send the data to a random address, it would still be received by the FPGA. The other way around was different, as the computer and Matlab ruth-lessly filter out all packets that are not explicitly addressed to them. That means that the FPGA had to set the headers just right. Moreover, IP protocol requires a valid checksum, that is, a cyclic redundancy check, appended at the end of the packet, so that the receiving hardware can verify that the data is not corrupted. If the checksum does not pass the test, the packet is dropped. That is, the FPGA design had to compute the CRC-32 for each data frame. All these problems were resolved, but required a certain effort.

At 1 Gbit/s Ethernet data transfer rate, sending the acquired signal from the FPGA to the computer requires a fraction of a millisecond. However, Matlab needs a minimal 0.6 s delay to process this data. I could not find the reasons for such slow buffering, and we had no choice but to live with it. But it is important to note that, should this Matlab problem be resolved, each iteration of the experiment could, in principle, run several orders of magnitude faster.

The second big challenge, encountered and solved by Michiel Hermans, was already described in Sect. 3.5.5. At first, the error rate, similar to the one plotted in Fig. 3.8, did not want to decrease below 0.2. We fought with the settings of the intensity modulator for a couple of days, trying to make it as symmetric as possible, untill Michiel Hermans came up with a much better solution. Adding a third "blanc" pass increased the experiment runtime by 33%, but allowed us to achieve much better results with the setup.

## 3.7 Conclusion

Our work confirms the results anticipated in [5, 6]: the performance of delay-based reservoir computers can be drastically improved by optimising both input and output masks. Furthermore, following the proposal of [7], we showed that the underlying hardware is capable of running a large part of its own optimisation process. We performed our demonstrations on a fast electro-optical system (whose speed could be readily improved by several orders of magnitude, see, e.g. [20]), and on tasks considered hard in the RC community. Importantly, our work has revealed that the BP algorithm is robust against various experimental imperfections, as the performance gains we obtained on all three tasks were similar to those predicted by numerical simulations.

Although our experiment relies on the sine nonlinearity and its cosine deriva-tive, other nonlinear functions can also be successfully realised in hardware with their derivatives. For instance, the so-called linear rectifier function, which trun-cates the input signal below a certain threshold, is a popular activation function in neural architectures [21]. Its derivative is a simple binary function which can be easily implemented using an analogue switch, as in [7]. In [22] it is shown how to implement a sigmoid nonlinearity and its derivative. And in [16, 23] the nonlinearity is quadratic, and therefore the derivative, which is linear, should also be easy to implement. Furthermore, the BP algorithm is robust against imperfect implementation of the derivative, as shown here in Sect. 3.5.5 and also in the Supplementary Material of [7] (see Supplementary Note 4). Therefore we expect that physical implementation of the BP algorithm will be possible in a wide variety of physical systems.

Our setup still requires some slow digital processing to perform the masking and to compute gradients from the recorded signals. Performing masking operations in analogue hardware, however, is actively being researched: several readout schemes have been proposed [24–26] and a few have been realised experimentally [24, 25]. Moreover, Chap. 5 of this thesis presents a numerical experiment demonstrating how to improve analogue readout layers with online learning. These approaches could be used to speed up the present setup. Another limitation is the large amount of data transfered between the FPGA and the computer. Implementing the full training algorithm on the FPGA would drastically increase the speed of the experiment.

Nowadays, there is an increased interest in unconventional, neuromorphic computing, as this could allow for energy efficient computing, and may provide a solution to the predicted end of Moore's law [27]. These novel approaches to computing will likely be made with components that exhibit strong element-to-element variability, or whose characteristics evolve slowly with time. Self-learning hardware may be the solution that enables these systems to fulfil their potential. The results in [7] and in this work therefore constitute an important step towards this goal.

Contrary to my first experiment, this project was completed in a couple of months. Sadly, we could not develop this research direction any further as Michiel Hermans left the team shortly after our paper [1] was written.

# References

1. Hermans, Michiel, Piotr Antonik, Marc Haelterman, and Serge Massar. 2016. Embodiment of learning in electro-optical signal processors. *Physical Review Letters* 117: 128301.
2. Rumelhart, David E., James L. McClelland, and PDP Research Group. 1986. Parallel distributed processing: explorations in the microstructure of cognition. In *Learning internal representations by error propagation*, vol. 1, 318–362. Cambridge, MA, USA: MIT Press.
3. Werbos, Paul J. 1988. Generalization of backpropagation with application to a recurrent gas market model. *Neural Networks* 1 (4): 339–356.
4. LeCun, Yann, Yoshua Bengio, and Geoffrey Hinton. 2015. Deep learning. *Nature* 521 (7553): 436–444.
5. Hermans, Michiel, Joni Dambre, and Peter Bienstman. 2015. Optoelectronic systems trained with backpropagation through time. *IEEE Transactions on Neural Networks and Learning Systems* 26 (7): 1545–1550.
6. Hermans, Michiel, Miguel Soriano, Joni Dambre, Peter Bienstman, and Ingo Fischer. 2015. Photonic delay systems as machine learning implementations. *JMLR* 16: 2081–2097.
7. Hermans, Michiel, Michaël Burm, Thomas Van Vaerenbergh, Joni Dambre, and Peter Bienstman. 2015. Trainable hardware for dynamical computing using error backpropagation through physical media. *Nature Communications* 6: 6729.

8. Ikeda, Kensuke, and Kenji Matsumoto. 1987. High-dimensional chaotic behavior in systems with time-delayed feedback. *Physica D: Non-Linear Phenomena* 29 (1): 223–235.

9. Nesterov, Yurii. 1983. A method of solving a convex programming problem with convergence rate O (1/k2). *Soviet Mathematics Doklady* 27 (2): 372–376.

10. Sutskever, Ilya, James Martens, George Dahl, and Geoffrey Hinton. 2013. On the importance of initialization and momentum in deep learning. In *Proceedings of the 30th international conference on machine learning (ICML-13)*, 1139–1147.

11. Paquot Yvan, François Duport, Anteo Smerieri, Joni Dambre, Benjamin Schrauwen, Marc Haelterman, and Serge Massar. 2012. Optoelectronic reservoir computing. *Scientific Reports* 2: 287.

12. Zimmermann, Hubert. 1980. OSI reference model–the ISO model of architecture for open systems interconnection. *IEEE Transactions on Communications* 28 (4): 425–432.

13. Garofolo, John, S., and NIST. 1993. *TIMIT Acoustic-phonetic continuous speech corpus*. Linguistic Data Consortium.

14. Triefenbach, Fabian, Azarakhsh Jalalvand, Benjamin Schrauwen, and Jean-Pierre Martens. 2010. Phoneme recognition with large hierarchical reservoirs. *Advances in Neural Information Processing Systems* 23: 2307–2315.

15. Triefenbach, Fabian, Kris Demuynck, and Jean-Pierre Martens. 2014. Large vocabulary continuous speech recognition with reservoir-based acoustic models. *IEEE Signal Processing Letters* 21 (3): 311–315.

16. Vinckier, Quentin, François Duport, Anteo Smerieri, Kristof Vandoorne, Peter Bienstman, Marc Haelterman, and Serge Massar. 2015. High-performance photonic reservoir computer based on a coherently driven passive cavity. *Optica* 2 (5): 438–446.

17. Hermans, Michiel, and Benjamin Schrauwen. 2012. Infinite sparse threshold unit networks. In *Proceedings of the international conference on artificial neural networks*, 612–619.

18. Bishop, Christopher M. 2006. *Pattern recognition and machine learning*. Springer.

19. Singh, Satnam. 2011. *Using the Virtex-6 Embedded Tri-Mode Ethernet MAC Wrapper v1.4 with the ML605 Board*. http://blogs.msdn.microsoft.com/satnam_singh/2011/02/11/using-the-virtex-6-embedded-tri-mode-ethernet-mac-wrapper-v1-4-with-the-ml605-board/.

20. Brunner, Daniel, Miguel C Soriano, Claudio R Mirasso, and Ingo Fischer. 2013. Parallel photonic information processing at gigabyte per second data rates using transient states. *Nature Communications* 4: 1364.

21. Glorot, Xavier, Antoine Bordes, and Yoshua Bengio. 2011. Deep sparse rectifier neural networks. In *14th international conference on artificial intelligence and statistics*, vol. 15 (106), p. 275.

22. Shi, Bingxue, and Chun Lu. 2002. *Generator of neuron transfer function and its derivative*. US Patent 6429699.

23. Vandoorne, Kristof, Pauline Mechet, Thomas Van Vaerenbergh, Martin Fiers, Geert Morthier, David Verstraeten, Benjamin Schrauwen, Joni Dambre, and Peter Bienstman. 2014. Experimental demonstration of reservoir computing on a silicon photonics chip. *Nature Communications* 5: 3541.

24. Smerieri, Anteo, François Duport, Yvan Paquot, Benjamin Schrauwen, Marc Haelterman, and Serge Massar. 2012. Analog readout for optical reservoir computers. In *Advances in neural information processing systems*, 944–952.

25. Duport, François, Anteo Smerieri, Akram Akrout, Marc Haelterman, and Serge Massar. 2016. Fully analogue photonic reservoir computer. *Scientific Reports* 6: 22381.

26. Vinckier, Quentin, Arno Bouwens, Marc Haelterman, and Serge Massar. 2016. Autonomous all-photonic processor based on reservoir computing paradigm. In *Conference on lasers and electro-optics*. Optical society of America. SF1F.1.

27. Waldrop, M.Mitchell. 2016. The chips are down for Moore's law. *Nature* 530: 144–147.

# Chapter 4
# Photonic Reservoir Computer with Output Feedback

This chapter presents, arguably, the most important experiment of my PhD. Not because I worked on it for a long time, but because it brought the most interesting, novel and even unexpected results.

Somewhat similar to Chap. 3, this project—as it is—was not part of my thesis. The original plan was to build an analogue output layer (see Chap. 5) and then feed the output signal back into the reservoir and see what it does. But how often do things go as planned, if ever? After the very good results obtained in the online training project (see Chap. 2), we decided to shortcut the analogue part of the experiment, for two reasons. First, an analogue layer would be more complex and we had no guarantee it would work sufficiently well. It is one thing to be able to compute the output of the system, but another to obtain a high-quality output signal that could be fed back as input. Second, analogue layers have already been researched [1–3], while nobody has ever reported an experimental reservoir computer with output feedback so far, neither digital, nor analogue. Therefore, the experimental demonstration of the novel features of such a system seemed a much more attractive and innovative idea.

The content of this chapter is based on our paper [4]. It is relatively long—we tested the setup on four new tasks (not yet considered by our lab) and performed an in-depth study of the reservoir output properties—and so is this chapter. Some additional results from our paper [5] are also included.

## 4.1 Introduction

Forecasting is one of the primary problems in science: how can one predict the future from the past? Over the past few decades, artificial neural networks have gained a significant recognition in the time series forecasting community. Similarly to previously employed statistics-based techniques, they are both data driven and non-linear. Differently, they are more flexible and do not require an explicit model of

© Springer International Publishing AG, part of Springer Nature 2018
P. Antonik, *Application of FPGA to Real-Time Machine Learning*,
Springer Theses, https://doi.org/10.1007/978-3-319-91053-6_4

the underlying process. A review of artificial neural networks models for time series forecasting can be found in [6]. Reservoir computing can be readily applied to short-term prediction tasks, that focus on generating a few future timesteps. As for long-horizon forecasting, that involves predicting the time series for as long as possible, it requires a small modification of the architecture, namely by feeding the RC output signal back into the reservoir. This additional feedback significantly enriches the internal dynamics of the reservoir, enabling it to generate time series autonomously, i.e. without receiving any input signal. With this modification, reservoir computing can be used for long-term prediction of chaotic series [7–11]. In fact, to the best of our knowledge, this approach holds the record for such chaotic time series prediction [10, 11]. A reservoir computer with output feedback can also achieve the easier task of generating periodic signals [12–14], and sine waves with a tunable frequency [15–17].

The aim of the project described in this chapter is to explore these novel applications experimentally. As a matter of fact, they have been widely studied numerically, but no experimental implementation has been reported so far. There are several diverse motivations for this investigation. First of all, reservoir computing is a biologically inspired algorithm: one of the main motivations of the seminal paper [18] was to propose how microcircuits in the neocortex could process information. More recently, it has been realised that the cerebellum and a reservoir computer have a very similar structure [19, 20]. Generating time series with specific attributes is an important property of biological neural and chemical circuits (for e.g. movement control, biological rhythms). Are biological circuits similar to reservoir computers used to generate trainable time series? Experimental investigation of this process can shed light on this tantalising question, for example by clarifying which kinds of time series, and what training processes are robust to experimental imperfections.

Second, generation of time series with specific characteristics is an important task in signal generation and processing. Given the prospect that photonic reservoir computing could carry out ultra-fast and low energy optical signal processing, this is an important area to explore, again with the aim of understanding which tasks are robust to experimental imperfections.

Finally, this investigation raises a new fundamental question in nonlinear dynamics: for a system that emulates a known chaotic time series, how does one quantify the quality of the emulation. Answering this question becomes crucial in case of experimental implementations, as physical systems are affected by noise, and thus cannot do better that generating an approximate, noisy emulation of the target chaotic time series. The comparison techniques used previously in numerical investigations fail in such situations, and one needs to develop new evaluation metrics.

Experimental realisation of these ideas requires, in general, a readout layer fast enough to generate and feed the output signal back in real-time. Several analogue solutions have been investigated recently [1–3], but none are as yet capable of performing sufficiently well in this application. In fact, successful training of an analogue readout layer with offline learning methods, used in most experimental RC setups up to now, requires a very precise model of the readout setup, which is hardly achievable experimentally. As demonstrated in [2], it is virtually impossible to characterise each hardware component of the setup with sufficient level of accuracy. The reason for this difficulty is that the output is a weighted sum with positive and negative coefficients of the internal states of the reservoir. Therefore, errors in the coefficients can build up and become comparable to the value of the desired output. For this reason, we chose the approach of a real-time digital readout layer implemented on a FPGA chip. The use of high-speed dedicated electronics makes it possible to compute the output signal in real time, as has been demonstrated in Chap. 2, and feed it back into the reservoir. In order to keep the experiment simple, we used as reservoir the opto-electronic delay system introduced in [21–23] and discussed in Sect. 1.2.4, that has shown state-of-the-art results on several benchmark tasks and is fairly easy to operate.

We show that that our experimental reservoir computer can successfully solve two periodic time series generation tasks: frequency and random pattern generation, that have been previously investigated numerically [13, 15, 16]. The first task allows to reveal different timescales within the neural network, and the second can be used to quantify the memory of the reservoir. The photonic computer manages to generate both sine waves and random patterns with high stability (verified on the timescale of several days). Furthermore, we apply the RC to emulation of two chaotic attractors: Mackey-Glass [24] and Lorenz [25] systems. In the literature, the emulation performance on these tasks is quantified in terms of the prediction horizon, i.e. the duration for which the RC can accurately follow a given trajectory on the chaotic attractor [11]. However, this method is not applicable in the presence of a relatively high level of experimental noise, with a Signal-to-Noise Ratio (SNR) of roughly 40 dB, as will be discussed in Sect. 4.7.1. This noise was not problematic in previous experiments using the same opto-electronic reservoir [22, 26], but turns out to be intolerable for a system with output feedback. This raises the question of how to evaluate a system that emulates a known chaotic time series in the presence of noise. In this study, we introduce several new approaches, such as frequency spectrum comparison and randomness tests. These approaches are based on well-known signal analysis techniques. To the best of our knowledge, they are employed for the first time here for the evaluation of a chaotic signal generated by a reservoir computer. Our results show that, although the RC struggles at following the target trajectory on the chaotic attractor, its output accurately reproduces the core characteristics of the target time series.

## 4.2 Reservoir Computing with Output Feedback

The introduction of the output feedback requires a minor change of notations, used in Sect. 1.1.3. Since the RC can now receive two different signals as input, we shall denote $I(n)$ the input signal, which can be either the external input signal $I(n) = u(n)$, or its own output, delayed by one timestep $I(n) = y(n-1)$.

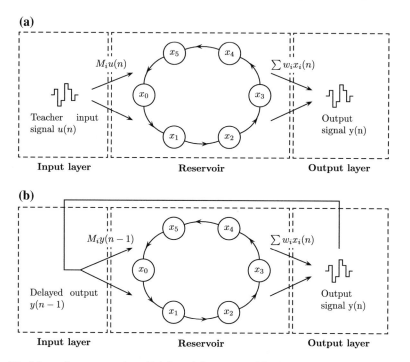

**Fig. 4.1** Schematic representation of (**a**) the training stage and (**b**) the autonomous run of a reservoir computer. For simplicity, a small network with $N = 6$ nodes is depicted. During the training phase, the reservoir is driven by a teacher input signal $u(n)$, and the readout weights $w_i$ are optimised for the output $y(n)$ to match $u(n + 1)$ as accurately as possible. During the autonomous run, the teacher signal $u(n)$ is switched off and the reservoir is driven by its own output signal $y(n)$. The readout weights $w_i$ are fixed and the performance of the system is measured in terms of how long or how well it can generate the desired output

The reservoir computer is operated in two stages, depicted in Fig. 4.1: a training phase and an autonomous run. During the training phase, the reservoir computer is driven by a time-multiplexed teacher signal $I(n) = u(n)$, and the resulting states of the internal variables $x_i(n)$ are recorded. The teacher signal depends on the task under investigation (which will be introduced in Sect. 4.3). The system is trained to predict the next value of the teacher time series from the current one, that is, the readout weights $w_i$ are optimised so as to get as close as possible to $y(n) = u(n + 1)$. Then, the reservoir input is switched from the teacher sequence to the reservoir output signal $I(n) = y(n - 1)$, and the system is left running autonomously. The reservoir output $y(n)$ is used to evaluate the performance of the experiment.

## 4.3   Time Series Generation Tasks

The output feedback allows the computer to generate time series autonomously, that is, without any external input. We tested the capacity of the experiment to produce both periodic and chaotic signals, with two tasks in each category.

### 4.3.1   Frequency Generation

Frequency generation is the simplest time series generation task considered here. The system is trained to generate a sine wave given by

$$u(n) = \sin(\nu n), \tag{4.1}$$

where $\nu$ is a real-valued relative frequency. The physical frequency $f$ of the sine wave depends on the experimental roundtrip time $T$ (see Sect. 1.2.4) as follows

$$f = \frac{\nu}{2\pi T}. \tag{4.2}$$

This task allows to measure the bandwidth of the system and investigate different timescales within the neural network.

### 4.3.2   Random Pattern Generation

A natural step forward from the frequency generation task is random pattern generation. Instead of a regularly-shaped continuous function, the system is trained to generate an arbitrarily-shaped discontinuous function (that remains periodic, though). In this context, a pattern is a sequence of $L$ randomly chosen real numbers (here within the interval $[-0.5, 0.5]$) that is repeated periodically to form an infinite periodic time series [8]. Similarly to the physical frequency in Sect. 4.3.1, the physical period of the pattern is given by $\tau_{\text{pattern}} = L \cdot T$. The aim is to obtain a stable pattern generator, that reproduces precisely the pattern and does not deviate to another periodic behaviour. To evaluate the performance of the RC, we compute the MSE between the reservoir output signal and the target pattern signal during both the training phase and the autonomous run, and set the maximal threshold to $10^{-3}$. This value is somewhat arbitrary, and one could have picked a different threshold. As will be illustrated in Figs. 4.12 and 4.16 in Sect. 4.7, the $10^{-3}$ level corresponds to the point where the RC strongly deviates from the starting trajectory on the chaotic attractor. For consistency, we have used this threshold in all our experiments, for all tasks. If the error does not grow above the threshold during the autonomous run, the system is considered to accurately generate the target pattern. We also tested the long-term stability on several patterns by running the system for several hours, as will be described in Sect. 4.7.

### 4.3.3  Mackey-Glass Chaotic Series Prediction

The Mackey-Glass delay differential equation

$$\frac{dx}{dt} = \beta \frac{x(t-\tau)}{1+x^n(t-\tau)} - \gamma x, \tag{4.3}$$

with $\tau, \gamma, \beta, n > 0$ was introduced to illustrate the appearance of complex dynamics in physiological control systems [24]. To obtain chaotic dynamics, we set the parameters as in [11]: $\beta = 0.2, \gamma = 0.1, \tau = 17$ and $n = 10$. In these conditions, the Kaplan-Yorke dimension of the chaotic attractor is 2.1 [27].

The equation was solved using the Runge-Kutta 4 method [28] with a stepsize of 1.0. To avoid repeating computations, we pre-generated a sequence of $10^6$ samples that we used for all numerical and experimental investigations.

The MSE is used to evaluate both the training phase and the autonomous run. During the latter, the system does not receive the the correct teacher signal anymore, and thus slowly deviates from the desired trajectory. Therefore, we compute the number of correct prediction steps, i.e. steps for which the MSE stays below the $10^{-3}$ threshold (see Sect. 4.3.2), and use this figure to evaluate the performance of the system.

### 4.3.4  Lorenz Chaotic Series Prediction

The Lorenz equations, a system of three ordinary differential equations

$$\frac{dx}{dt} = \sigma(y-x), \tag{4.4a}$$

$$\frac{dy}{dt} = -xz + rx - y, \tag{4.4b}$$

$$\frac{dz}{dt} = xy - bz, \tag{4.4c}$$

with $\sigma, r, b > 0$, was introduced as a simple model for atmospheric convection [25]. The system exhibits chaotic behaviour for $\sigma = 10, b = 8/3$ and $r = 28$ [29], that we used in this study. These parameters yield a chaotic attractor with the highest Lyapunov exponent of $\lambda = 0.906$ [11]. The system was solved using Matlab's `ode45` solver and a stepsize of 0.02, as in [11]. We used all computed points, meaning that one timestep of the reservoir computer corresponds to a step of 0.02 in the Lorenz time scale. To avoid unnecessary computations and save time we pre-generated a sequence of $10^5$ samples that we used for all numerical and experimental investigations. Following [11], we used the $x$-coordinate trajectory for training and testing, that we scaled by a factor of 0.01.

## 4.4 Experimental Setup

Our experimental setup, schematised in Fig. 4.2, consists of two main components: the opto-electronic reservoir and the FPGA board. The structure and operation of the opto-electronic reservoir have been discussed in Sect. 1.2.4. In this section we will focus on a few particular aspects of this experiment. The functioning of the FPGA will be presented in Sect. 4.5.

With time-multiplexed neurons, the maximal reservoir size is imposed by the ratio between the delay from the fibre spool (Spool) and the sampling frequency of the ADC. While increasing the latter involves relatively high costs, one can lengthen the delay line fairly easily. In this work, we used two spools of single mode fibre with approximate lengths of 1.6 and 10 km. The first spool, identical to the previous experiments, described in Chaps. 2 and 3, produced a delay of 7.93 μs and could take in $N = 100$ neurons. The second spool was used to increase the delay up to 49.2 μs and the reservoir size up to $N = 600$. In both cases, the reservoir states were sampled at approximately 200 MHz. The precise frequency depends on the delay loop and will be given in Sect. 4.5. Each state was averaged over 16 samples in order to decrease the noise and remove the transients induced by the finite bandwidth of the Digital-to-Analogue converter (DAC).

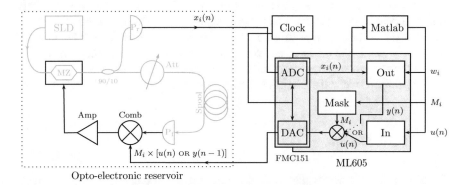

**Fig. 4.2** Schematic representation of the experimental setup. The opto-electronic reservoir has been introduced in Sect. 1.2.4. The FPGA board implements the readout layer and computes the output signal $y(n)$ in real time. It also generates the analogue input signal $I(n)$ and acquires the reservoir states $x_i(n)$. The computer, running Matlab, controls the devices, performs the offline training and uploads all the data ($u(n)$, $w_i$ and $M_i$) on the FPGA

The experiment is operated as follows. First, the input mask $M_i$ and the teacher signal $u(n)$ are generated in Matlab and uploaded on the FPGA board. The latter generates the masked input signal $M_i \times u(n)$ and sends it to the reservoir via the DAC. The resulting reservoir states $x_i(n)$ are sampled, averaged, and transferred to the computer in real time by the FPGA. That is, the design uses minimal memory (a small buffer for the Ethernet frames) and thus allows to capture the reservoir states without limitation of the time interval. After acquisition of the desired amount of data (reservoir states) in Matlab, the optimal readout weights $w_i$ are computed and uploaded on the FPGA board. Because of the relatively long delay (compared to the microsecond timescale of the experiment) needed for the offline training, the reservoir needs to be reinitialised in order to restore the desired dynamics of the internal states prior to running it autonomously. To this end, we drive the system with an initialisation sequence of 128 timesteps (as illustrated in Fig. 4.9), before coupling the output signal with the input and letting the reservoir computer run autonomously. In this stage, the FPGA computes the output signal $y(n)$ in real time, then creates a masked version $M_i \times y(n)$ and sends it to the reservoir via the DAC.

As the neurons are processed sequentially, the output signal $y(n)$ can only be computed in time to update the 24-th neuron $x_{23}(n+1)$. In other words, the first 23 neurons do not "see" the input signal $I(n)$ because it can not be computed and delivered in time. Therefore, we set the first 23 elements of the input mask $M_i$ to zero. That way, all neurons can contribute to solving the task, despite the first 23 lacking the input information. Note that this reflects an aspect that is inherent to any experimental time-multiplexed reservoir computer with output feedback. Indeed, the output $y(n)$ has to be computed after the acquisition of the last neuron $x_{N-1}(n)$ at timestep $n$, but before the first neuron $x_0(n+1)$ of the following timestep. However, in time-multiplexed implementation of reservoir computing, these states are consecutive, and the experiment cannot be paused to let $y(n)$ be computed. Therefore, there may be a delay, whose duration depends on the hardware used, before $y(n)$ is computed and can be fed back into the reservoir. In the present experiment, this delay is approximately 115 ns, which corresponds to 23 neuron durations. This delay is mainly due to propagation times between the intensity modulator (MZ) and the ADC in one hand, and the DAC and the resistive combiner on the other. The FPGA computation time also plays a role here, but it does not exceed 20 ns with our design (see Sect. 4.5). As we will see below this delay has an impact on system performance.

## 4.5  FPGA Design

The same Xilinx ML605 evaluation board (see Sect. 1.3.3) is used in this experiment, paired with a 4DSP FMC151 daughtercard. The simplified schematics of the design is depicted in Fig. 4.3. Rectangular boxes represent modules (entities), and rounded rectangles stand for electronic components on the ML605 board, namely the FMC151 daughtercard and the onboard Marvell Alaska PHY device (88E1111) for Ethernet communications (ETH).

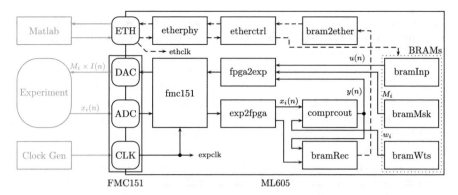

**Fig. 4.3** Simplified schematics of the FPGA design. Modules (entities) are represented by rectangular boxes, onboard electronic components are shown with rounded rectangles. External hardware, such as the computer, running Matlab, the opto-electronic reservoir and the external clock generator are shown in grey. The design is driven by two clocks: the experimental clock `expclk` and the Ethernet clock `ethclk`. Signals from these two clock domains are drawn in solid and dashed lines, respectively

The operation of the FPGA is controlled from the computer, running Matlab, via a simple custom protocol through a Gbit Ethernet connection. Data and various commands, such as memory read/write, or state change, are encapsulated into standard UDP packages. The `etherphy` module interfaces the FPGA design with the Marvell Ethernet PHY chip, and the `etherctrl` module receives the UDP packets (frames) and decodes the commands and the data. It also creates the frames for sending data from FPGA to the computer.

Blocks of Random-Access Memory (BRAM) are used to store data generated on the computer and uploaded on the FPGA: the teacher inputs $u(n)$, input masks $M_i$, and readout weights $w_i$. Each type of data is handled by a specific module, since they vary in size (e.g. 600 values for the input mask and up to 3000 for the teacher signal) and resolution, as will be explained below. The `bramRec` is a buffer-like module, designed to transfer the signal recorded from the experiment directly to the computer through Ethernet, without permanently storing it in memory. It consists of two blocks of RAM of 2048 bytes each, that are used as follows: while the recorded signal is written into the first block, `bram2ether` reads the contents of the second, that is then encapsulated into four UDP frames sent to the computer. When the first block is full, the blocks are switched and the process continues.

The FMC151 daughtercard is interfaced with the rest of the design through the `fmc151` module, that outputs two 14-bit signals from the ADCs and receives two 16-bit signals from the DACs. The FMC151 card is also used to deliver a clock signal from an external clock generator, that produces a high-precision signal, allowing to synchronise the FPGA with the delay loop of the experimental setup. This clock signal was generated by the Hewlett Packard 8648 A signal generator. As our experiment has two delay loops (see Sect. 4.4), these need to be precisely synchronised. To this end, we fine-tuned the clock frequency so as to fit 16 samples per neuron into the roundtrip time $T$. Specifically, for the large spool with $N = 600$ reservoir states, we sampled at 195.4472 MHz, while for the small spool with $N = 100$ neurons we sampled at 203.7831 MHz.

The fpga2exp module controls the signal sent to the opto-electronic reservoir through the DAC. During the training phase, it generates the masked input signal $M_i \times u(n)$ by multiplying the inputs $u(n)$ by the mask $M_i$, both being read from the BRAMs. During the autonomous run, it receives the reservoir output signal $y(n)$, computed by the comprcout module, masks it and transfers to the DAC.

The neuron states $x_i(n)$ from the photonic reservoir are sampled and averaged by the exp2fpga module. During the training phase, these are buffered in bramRec, then processed by the bram2ether module and sent to the computer. During the autonomous run, the reservoir states are used by the comprcout, together with the readout weights $w_i$, read from the bramWts memory, to compute the reservoir output $y(n)$. It is then injected back into the reservoir through the fpga2exp, and also transferred to the computer through the bramRec and bram2ether modules.

The design is driven by two clocks: the experimental clock expclk (around 200 MHz, depending on the loop delay $T$, see above) that operates data acquisition and generation modules and allows to synchronise the FPGA with the experiment, and the 125 MHz Ethernet clock ethclk. Both clocks have to be managed properly within the design, as several signals, such as inputs or weights, coexist in both clock domains. That is, data to BRAMs comes from the Ethernet modules, and is thus driven by the ethclk clock. On the other hand, this same data is used by the fpga2exp module, and has to appear in the expclk clock domain. To this end, we exploit the dual-port capability of Xilinx block RAMs. That is, data is written into memory blocks through port A at clock ethclk and read from port B at clock expclk (and vice versa for the bramRec). This allows for smooth transition of data between clock domains. The two clock domains are depicted in Fig. 4.3 as follows: signals running at expclk are shown in solid lines, and those clocking at ethclk are drawn with dashed lines.

Similarly to the design in Sect. 2.5, care should be taken with bit-represen-tation of real numbers. In this experiment, the main constraint comes from the ADC and DAC, limited to 14 and 16 bits, respectively. Numerical simulations, reported in [8], show that such precision is sufficient for all tasks studied in this work. Furthermore, it was shown that the precision of the readout weights $w_i$ has a significant impact on the performance of the system [8]. For this reason we designed the experiment for optimal utilisation of the available resolution. The reservoir states were tuned to lie within a $]-1, +1[$ interval. They are thus represented as 16-bit integers, with 1 bit for the sign and 15 bits for the decimal part. Another limitation comes from the DSP48E slices, used to multiply the states $x_i(n)$ by the readout weights $w_i$, and designed to multiply a 25-bit integer by a 18-bit integer (see Sect. 1.3.3). To meet these requirements, we keep the readout weights $w_i$ within the $]-1, 1[$ interval and represent them as 25-bit integers, with 1 sign bit and 24 decimal bits. To ensure that $w_i \in ]-1, 1[$, we amplify the reservoir states digitally inside the FPGA. That is, the $x_i(n)$ are multiplied by 8 after acquisition, prior to computing the output signal $y(n)$.

## 4.6  Numerical Simulations

In addition to the physical experiments, we investigated the proposed setup in numerical simulations, to have a point of comparison and identify possible malfunctions. To this end, we developed three models that simulate the experiment to different degrees of accuracy. Our custom Matlab scripts are based on [8, 22].

**Idealised model** It incorporates the core theoretical characteristics of our reservoir computer: the ring-like architecture, the sine nonlinearity, and the linear readout layer (as described by Eqs. 1.5 and 1.6). All experimental considerations are disregarded. We use this model to define the maximal performance achievable in each configuration.

**Noiseless experimental model** This model emulates the most influential features of the experimental setup: the high-pass filter of the amplifier, the finite resolution of the ADC and DAC, and precise input and feedback gains. This model allows to cross-check the experimental results and to identify the problematic points. That is, if the experiment performs much worse than the model, then, most likely, something does not work as it should.

**Noisy experimental model** Physical experiments come in pairs with noise, which, as will be explained below in Sect. 4.7, has a significant impact on performance. To compare our experimental results with a more realistic model, we estimated the level of noise present in the experimental system (see Sect. 4.7.1), and incorporated this noise into the noisy version of the experimental model.

## 4.7  Results

In this section we present the experimental results, compare them to numerical simulations and discuss the performance of the reservoir computer on each task introduced in Sect. 4.3.

The two periodic signal generation tasks were solved using a small reservoir with $N = 100$ and a fibre spool of approximately 1.6 km. The chaotic signal generation tasks, being more complex, required a large reservoir of $N = 600$ for sufficiently good results, that we fit in a delay line of roughly 10 km.

### 4.7.1  Noisy Reservoir

For most tasks studied here, we found the experimental noise to be the major source of performance degradation in comparison to previously reported numerical investigations [8]. This disparity stems from an ideal noiseless reservoir considered in [8], while our experiment is noisy. This noise can come from the active and even passive components of the setup: the amplifier, which has a relatively high gain and is therefore very sensitive to small parasitic signals (e.g. from the power source), the DAC, the photodiodes and the optical attenuator (shot noise). In-depth experimental investigations have shown that, in fact, each component contributes more or less equally to the overall noise level. Thus, it cannot be reduced by replacing one "faulty" component. Neither can it be averaged out, as the output value has to be computed at each timestep. This noise was found to have a marked impact on the results, as will be shown in the following sections. To further investigate this issue, we estimated the level of noise present in the experimental system and incorporated it to the numerical models. This allows us to examine different levels of noise, and even switch it "off" completely, which is impossible experimentally.

Figure 4.4 shows numerical and experimental states of a reservoir with $N = 100$ neurons, as received by the readout photodiode. That is, the curves depict the time-multiplexed neurons: each point represents a reservoir state $x_{0...99}(n)$ at times $n = 1$ and $n = 2$. The system does not receive any input signal $I(n) = 0$. The experimental signal is plotted with a solid grey line. We use it to compute the experimental noise level by taking the standard deviation of the signal, which gives $2 \times 10^{-3}$. We then replicate this noise level in the noisy experimental model to compare experimental results to numerical simulations. The dotted black curve in Fig. 4.4 shows the response of the noisy experimental model, with the same amount of Gaussian noise (standard deviation of $2.0 \times 10^{-3}$) as in the experiment. The choice of a Gaussian noise distribution was validated by experimental measurements.

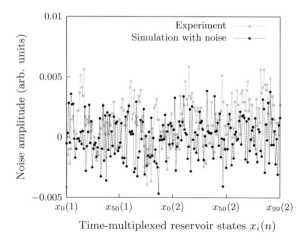

**Fig. 4.4** Illustration of the noisiness of the experimental reservoir. Experimental (solid grey line) and numerical (dotted black line) reservoir states $x_i(n)$ are scaled so that in normal experimental conditions (non-zero input) they would lie in a $[-1, 1]$ interval. Despite the null input signal $I(n) = 0$, the actual neurons are non-zero because of noise. Numerical noise was generated with a Gaussian random distribution with standard deviation of $1 \times 10^{-3}$ so that to reproduce the noise level of the experiment

The experimental noise level can also be characterised by the Signal-to-Noise Ratio (SNR), defined as [30]

$$\text{SNR} = 10 \log_{10} \left( \frac{\text{RMS}^2_{\text{signal}}}{\text{RMS}^2_{\text{noise}}} \right),$$

where RMS is the Root Mean Square value, given by

$$\text{RMS}(x_i) = \sqrt{\frac{1}{N} \sum_{i=1}^{N} x_i^2}.$$

We measured $\text{RMS}_{\text{signal}} = 0.2468$ and $\text{RMS}_{\text{noise}} = 0.0023$, so the SNR is equal to approximately 40 dB in this case. Note that this figure is given as an indicator of order of magnitude only as the RMS of the reservoir states depends on the gain parameters ($\alpha$ and $\beta$ in Eqs. 1.5) and varies from one experiment to another.

### 4.7.2  Frequency Generation

We found the frequency generation task to be the only one not affected by noise: our experimental results matched accurately the numerical predictions reported in [16]. From this study, we expected a bandwidth of $\nu \in [0.06, \pi]$ with a 100-neuron reservoir. The upper limit is a signal oscillating between $-1$ and $1$ and is given by half of the sampling rate of the system (the Nyquist frequency [31]). The lower limit is due to the reservoir memory. In fact, low-frequency oscillations correspond to longer periods, and the neural network can no longer "remember" a sufficiently long segment of the sine wave so as to keep generating a sinusoidal output. These numerical results are confirmed experimentally here.

We tested our setup on frequencies $\nu$ ranging from 0.01 to $\pi$. We found that frequencies within the $[0.1, \pi]$ interval are generate accurately with any random input mask. Lower frequencies between 0.01 and 0.1, on the other hand, were produced properly with some random masks, but not all. Since this is where the lower limit of the bandwidth lies, we investigated the $[0.01, 0.1]$ interval more precisely. For each frequency, we ran the experiment 10 times for $10^4$ timesteps with different random input masks and counted the number of times the reservoir produced a sine wave with the desired frequency (MSE $< 10^{-3}$, see Sect. 4.3.2) and amplitude of 1. The results are shown in Fig. 4.5. Frequencies below 0.05 are not generated correctly with most input masks. At $\nu = 0.7$ the output is correct most of the times, and for $\nu = 0.08$ and above the output sine wave is correct with any input mask. Thus, we can conclude that the bandwidth of this experimental RC is $\nu \in [0.08, \pi]$. Considering the roundtrip time $T = 7.93 \, \mu s$, this results in a physical bandwidth of 1.5–63 kHz. Note that frequencies within this interval can be generated with any random input

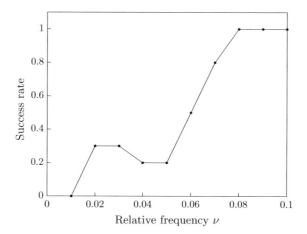

**Fig. 4.5** Verification of the lower limit of the reservoir computer bandwidth on the frequency generation task. Frequencies above 0.1 (not shown on the plot) are generated very well with any of the 10 random input mask. Frequencies below 0.05 fail with most input masks. We thus consider 0.08 as the lower limit of the bandwidth, but keep in mind that frequencies as low as 0.02 could also be generated, but only with a carefully picked input mask

**Fig. 4.6** Example of an autonomous run output signal for frequency generation task with $\nu = 0.1$. The experiment continues beyond the range of the figure

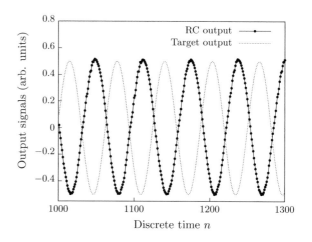

mask $M_i$. Lower frequencies, down to 0.02, could also be generated after choosing a suitable mask.

Figure 4.6 shows an example of the output signal during the autonomous run. The system was trained for 1000 timesteps to generate a frequency of $\nu = 0.1$, and successfully accomplished this task with a MSE of $5.6 \times 10^{-9}$.

These results were obtained by scanning the input gain $\beta$ and the feedback gain $\alpha$ to obtain the best results. It was found that $\beta$ has little impact on the system performance so long as it is chosen in the interval $\beta \in [0.02, 0.5]$, while the feedback gain $\alpha$, on the contrary, has to lie within a narrow interval of $\alpha \in [4.25, 5.25]$ dB

**Table 4.1** Optimal RC parameters for frequency generation

| $\alpha$ | $\beta$ | $V_\phi$ |
|---|---|---|
| 4.25–5.25 dB | 0.02–0.5 | 0.9 V |

(this corresponds approximately to $\alpha \in [0.85, 0.95]$), otherwise the reservoir yields very poor results. The DC bias $V_\phi$ of the MZ modulator was set to 0.9 V to ensure a symmetric transfer function ($\phi = 0$). These parameters are summarised in Table 4.1.

### 4.7.3 Random Pattern Generation

The random pattern generation task is more complex than frequency generation and is slightly affected by the experimental noise. The goal of this task is two-fold: "remember" a pattern of a given length $L$ and reproduce it for an unlimited duration. We have shown numerically that a noiseless reservoir with $N = 51$ neurons is capable of generating patterns up to 51-element long [8]. This is a logical result, as, intuitively, each neuron of the system is expected to "memorise" one value of the pattern. Simulations of a noisy reservoir with $N = 100$ neurons, similar to the experimental setup, show that the maximum pattern length is reduced down to $L = 13$. That is, the noise significantly reduces the effective memory of the system. In fact, the noisy neural network has to take into account the slight deviations of the output from the target pattern so as to be able to follow the pattern disregarding these imperfections. Figure 4.7 illustrates the manifestation of noise. Periodic oscillations of one neuron of the reservoir are shown, with intended focus on the upper values and

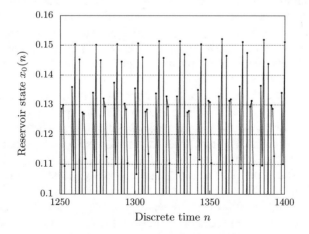

**Fig. 4.7** Example of behaviour of one neuron in a noisy experimental reservoir. For clarity, the range of the Y axis is limited to the area of interest. Because of noise, despite a periodic input signal $u(n)$, the reservoir state takes similar, but not identical values

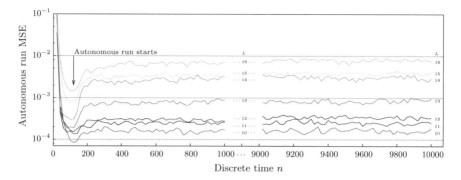

**Fig. 4.8** Evolution of MSE($n$) during experimental autonomous generation of periodic random patterns of lengths $L = 10, \ldots, 16$. The autonomous run starts at $n = 128$, as indicated by the arrow. Patterns shorter than 13 are reproduced with low MSE $< 10^{-3}$. Patterns longer than 14 are not generated correctly with MSE $> 10^{-3}$. In the latter case, the reservoir dynamics remains stable and periodic, but the output only remotely resembles the target pattern (see Figs. 4.9 and 4.10 for illustration)

an adequate magnification so as to see the small variations. The plot shows that the neuron oscillates between similar, but not identical values. This makes the generation task much more complex and requires more memory, which, in turn, shortens the maximal pattern length.

We obtained similar results in the experiments. Figure 4.8 shows the evolution of the MSE measured during the first 1000 timesteps of $10^4$-timestep autonomous runs with different pattern lengths. Plotted curves are averaged over 100 runs of the experiment, with 5 random input masks and 20 random patterns for each length. The initial minimum (at $n = 128$) corresponds to the initialisation of the reservoir (see Sect. 4.4), then the output is coupled back and the system runs autonomously. Patterns with $L = 12$ or less are generated very well and the error stays low. Patterns of length 13 show an increase in MSE, but they are still generated reasonably well. For longer patterns, the system deviates to a different periodic behaviour, and the error grows above $10^{-3}$.

Figure 4.9 shows an example of the output signal during the autonomous run. The system was trained for 1000 timesteps to generate a pattern of length 10. The reservoir computer successfully learned the desired pattern and the output accurately matches the target signal. Figure 4.10 illustrates a case with a longer patter ($L = 14$), that could not be learned by the system. As can be seen from the plot, the RC captured the general shape of the pattern, but cannot accurately generate individual points. The MSE of this run is $5.2 \times 10^{-3}$, which is above the acceptable $10^{-3}$ threshold.

We also tested the stability of the generator by running it for several hours (roughly $10^9$ timesteps) with random patterns of lengths 10, 11 and 12. The output signal, observed on a scope, remained stable and accurate through the whole test.[1]

---

[1]More precisely, every time I came to the lab to check.

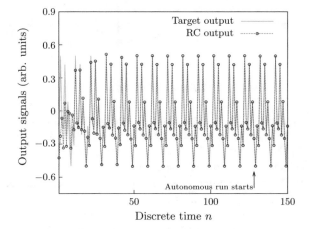

**Fig. 4.9** Example of an output signal for random pattern generation task, with a pattern of length 10. The reservoir is first driven by the desired signal for 128 timesteps (see Sect. 4.5), and then the input is connected to the output. Note that in this example the reservoir output requires about 50 timesteps to match the driver signal. The autonomous run continues beyond the scope of the figure

**Fig. 4.10** Example of an autonomous run output after 1950 timesteps, with a pattern of length $L = 14$. The RC outputs a periodic signal that clearly does not match the target pattern (MSE $= 5.2 \times 10^{-3}$)

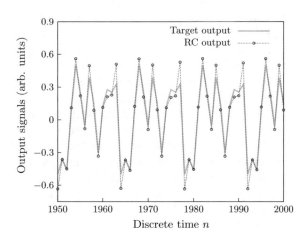

The above results were obtained by scanning the input gain $\beta$ and the feedback gain $\alpha$ to obtain the best results. As for frequency generation, it was found that $\beta$ has little impact on the system performance so long as it is chosen in the interval $\beta \in [0.1, 1]$, while the feedback gain $\alpha$, on the contrary, has to lie within a narrow interval of $\alpha \in [4.25, 5.25]$ dB (this corresponds approximately to $\alpha \in [0.85, 0.95]$). The DC bias of the MZ modulator was set to $V_\phi = 0.9$ V to ensure a symmetric transfer function ($\phi = 0$). These parameters are summarised in Table 4.2.

**Table 4.2** Optimal RC parameters for pattern generation

| $\alpha$ | $\beta$ | $V_\phi$ |
|---|---|---|
| 4.25–5.25 dB | 0.02–0.5 | 0.9 V |

#### 4.7.3.1  Numerical Study of the Impact of Noise

Since the noise plays such an important role, we performed a series of numerical experiments with different levels of noise to find out to what extent it affects the performance of the computer.[2] We used the noisy model of the experiment with Gaussian white noise with zero mean and standard deviations ranging from $10^{-2}$ to as low as $10^{-8}$. These simulations allow to estimate the expected performance of the experiment for different levels of noise.

Figure 4.11 shows the maximum pattern length $L$ that the reservoir computer is able to generate for different levels of noise. The maximal length is determined using the $10^{-3}$ autonomous error threshold, as described in Sect. 4.3.2. That is, if the NMSE does not grow above $10^{-3}$ during the autonomous run, the reservoir computer is considered to have successfully generated the given pattern. For statistical purposes, we used 10 different random patterns for each length $L$, and only counted the cases where the system have succeeded in all 10 trials. The results show that the noise level of $10^{-8}$ is equivalent to an ideal noiseless reservoir. As the noise level increases, the memory capacity of the reservoir deteriorates. At a level of $10^{-3}$, the maximum pattern length is decreased down to 10, which matches the experimental results presented here. For higher noise levels the results are, obviously, even worse.

Overall, these results show what level of noise one should aim for in order to obtain a certain performance from an experimental reservoir computer with output feedback. Our experiments have confirmed the numerical results for the noise level of $10^{-3}$. In principle, one could double the maximal pattern length by carefully re-building the same experiment with low-noise components, namely a weaker amplifier and a low-$V_\pi$ intensity modulator, which would lower the noise to $10^{-4}$. Switching to a passive setup, such as the coherently driven cavity reported in [32], could potentially lower the noise down to $10^{-5}$ or even $10^{-6}$, with performance approaching the maximum memory capacity.

### 4.7.4  Mackey-Glass Series Prediction

Chaotic time series generation tasks were the most affected by the experimental noise. This is not surprising, since, by definition, chaotic systems are very sensitive to initial conditions, which are affected by noise. Reservoir computing was first applied to this class of tasks in [11]. The authors numerically investigated the capacity of the computer to follow a given trajectory in the phase space of the chaotic attractor. We

---

[2]These results have been published in [5].

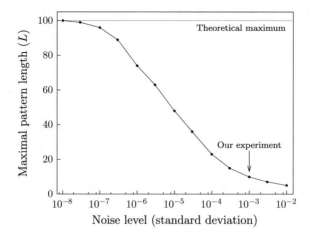

**Fig. 4.11** Impact of experimental noise on the performance of a reservoir computer with output feedback. The graph presents numerical results obtained with an accurate model of the experimental setup. Noise levels are shown as standard deviations of the Gaussian noise used in the simulations. The system was tested on the random pattern generation task and the performance metric is the maximal length $L$ of a pattern that the reservoir could generate. The theoretical maximum is $L = 100$, since we used a reservoir with $N = 100$ neurons. Noise levels of $10^{-8}$ and below are equivalent to an ideal noiseless system. The arrow indicates the experimental results presented in this work

**Table 4.3** Optimal RC parameters for the Mackey-Glass task

| $\alpha$ | $\beta$ | $V_\phi$ |
|---|---|---|
| 4.25–5.25 dB | 0.1–0.3 | 0.9 V |

also followed this approach at first, but since our experimental system performs as a "noisy" emulator of the chaotic attractor, its trajectory deviates very quickly from the target one, especially with a SNR as low as 40 dB (see Sect. 4.7.1). For this reason, we considered alternative methods to evaluate the performance of the system, as will be described below.

The system was trained over 1500 input samples and was running autonomously for 600 timesteps. In particular, we prepared 2100 steps of the Mackey-Glass series for each run of the experiment and used the first 1500 as a teacher signal $u(n)$ to train the system and the last 600 both as an initialisation sequence (see Sect. 4.5) and as a target signal $d(n)$ to compute the MSE of the output signal $y(n)$. These 2100 samples were taken from several starting points $t$ (see Eq. 4.3) in order to test the reservoir computer on different instances of the Mackey-Glass series. We scanned the input gain and the feedback attenuation ($\beta$ and $\alpha$ in Eqs. 1.5) to find optimal dynamics of the opto-electronic reservoir for this task. We used $\beta \in [0.1, 0.3]$ and tuned the optical attenuator in the range [4.25, 5.25] dB, which corresponds approximately to $\alpha \in [0.85, 0.95]$, with slightly different values for different instances of the Mackey-Glass series. The DC bias of the MZ modulator was set to $V_\phi = 0.9$ V to ensure a symmetric transfer function ($\phi = 0$). These parameters are summarised in Table 4.3.

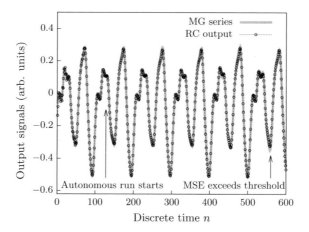

**Fig. 4.12** Example of reservoir computer output signal $y(n)$ (dotted black line) during autonomous run on the Mackey-Glass task. The system was driven by the target signal (solid grey line) for 128 timesteps and then left running autonomously, with $y(n)$ coupled to the input $I(n)$ (see Sect. 4.2). The MSE threshold was set to $10^{-3}$. The photonic reservoir computer with $N = 600$ was able to generate up to 435 correct values

**Fig. 4.13** Evolution of MSE during experimental autonomous generation of the Mackey-Glass chaotic time series (same run as in Fig. 4.12). The error curve, averaged over 200 timesteps, crosses the $10^{-3}$ threshold approximately between $n = 500$ and $n = 600$

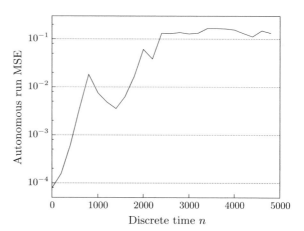

Figure 4.12 shows an example of the reservoir output $y(n)$ (dotted black line) during the autonomous run. The target Mackey-Glass series is shown in grey. The MSE threshold was set to $10^{-3}$ and the reservoir computer predicted 435 correct values in this example. Figure 4.13 displays the evolution of the MSE recorded during the same autonomous run. The plotted error curve was averaged over 200-timestep intervals. It exceeds the $10^{-3}$ threshold within $n \in [500, 600]$ and reaches a constant value of approximately $1.1 \times 10^{-1}$ after 2500 timesteps. At this point, the generated time series is completely off the target (see Fig. 4.14 for illustration).

The noise inside the opto-electronic reservoir, discussed in Sect. 4.7.1, makes the outcome of an experiment inconsistent. That is, repeating the experiment with

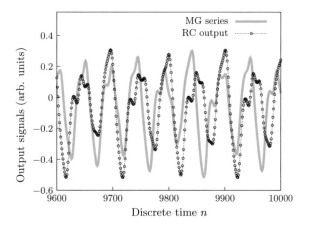

**Fig. 4.14** Output of the experimental reservoir computer (dotted black line) at the end of a long run of $10^4$ timesteps. Although the system does not follow the starting trajectory (solid grey line), its output still resembles visually the target time series

same parameters may result in significantly different prediction lengths. In fact, the impact of noise varies from one trial to another. In some cases it does not disturb the system much. But in most cases it induces a significant error on the output value $y(n)$, so that the neural network deviates from the target trajectory. To estimate the variability of the results, we performed 50 consecutive autonomous runs with the same readout weights and the same optimal experimental parameters. The system produced several very good predictions (of order of 400), but most of the outcomes were rather poor, with an average prediction length of 63.7 and a standard deviation of 65.2. We obtained similar behaviour with the noisy experimental model, using the same level of noise as in the experiments. Changing the ridge regression parameter in the training process (see Sect. 1.1.3) did not improve the results. This suggests that the reservoir computer emulates a "noisy" Mackey-Glass system, and therefore, using it to follow a given trajectory does not make much sense with such a high noise level. Nevertheless, the noise does not prevent the system from emulating the Mackey-Glass system—even if the output quickly deviates from the target, it still resembles the original time series. Therefore, we tried a few distinct methods of comparing the output of the system with the target time series.

We performed a new set of experiments, where, after a training phase of 1500 timesteps, the system was running autonomously for $10^4$ timesteps in order to collect enough points for data analysis. We then proceeded with a simple visual inspection of the generated time series, to check whether it still looks similar to the Mackey-Glass time series, and does not settle down to simple periodic oscillations. Figure 4.14 shows the output of the experimental reservoir computer at the end of the $10^4$-timestep autonomous run. It shows that the reservoir output is still similar to the target time series, that is, irregular and consisting of the same kind of uneven oscillations.

A more thorough way of comparing two time series that "look similar" is to compare their frequency spectra. Figure 4.15 shows the Fast Fourier Transforms of the original Mackey-Glass series (solid grey line) and the output of the experiment after a long run (dotted black line). Remarkably, the reservoir computer reproduces

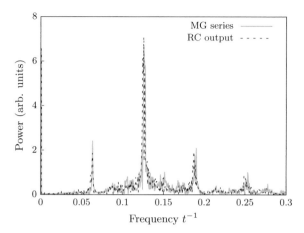

**Fig. 4.15** Comparison of Fast Fourier Transforms of the original Mackey-Glass series (solid grey line) and the time series generated by the photonic reservoir computer (dashed black line). The plot is limited to low frequencies as the power at higher frequencies is almost null. Dominant frequencies correspond to multiples of $1/\tau \approx 0.06$ (see Sect. 4.3.3). The experiment reproduces the target spectrum notably well

very accurately the spectrum of the chaotic time series, with its main frequency and several secondary frequencies.

Finally, we estimated the Lyapunov exponent of the generated time series, using the method described in the Supplementary Material of [11]. We obtained 0.01 for our experimental implementation. For the Mackey-Glass system considered here (see the parameters in Sect. 4.3.3), the value commonly found in the literature is 0.006. The slightly higher value of the Lyapunov exponent may simply reflect the presence of noise in the emulator.

### 4.7.5   Lorenz Series Prediction

This task was investigated in a similar way to the Mackey-Glass. The reservoir computer was trained over 3000 input samples and ran autonomously for 1000 timesteps. The 4000 samples were taken from an interval with even distribution of transitions between the two "wings" of the Lorenz attractor. In fact, we have noticed that the first 1000 samples of the sequence generated by the ode45 solver (see Sect. 4.3.4) contained more oscillations above zero than below, that is, a transient from the starting point to the actual chaotic attractor. This uneven distribution forced the reservoir computer to generate a biased output. We thus discarded the first 1000 values and trained the system over the interval [1000, 4000] (these initial transients were also removed in [11]). For optimal performance of the opto-electronic reservoir, we set the input gain to $\beta = 0.5$ and the feedback attenuation to $\alpha = 5.1$ dB. The DC bias

**Table 4.4** Optimal RC parameters for the Lorenz task

| $\alpha$ | $\beta$ | $V_\phi$ |
|---|---|---|
| 5.1 dB | 0.5 | 0.9 V |

of the MZ modulator was set to $V_\phi = 0.9$ V to ensure a symmetric transfer function ($\phi = 0$). These parameters are summarised in Table 4.4.

Figure 4.16 shows an example of the reservoir output $y(n)$ (dotted black line) during the autonomous run. The target Lorenz series is shown in grey. With the MSE threshold set to $10^{-3}$, the system predicted 122 correct steps, including two transitions between the wings of the attractor. As in the Mackey-Glass study, we performed 50 autonomous runs with identical parameters and readout weights and obtained an average prediction horizon of 46.0 timesteps with a standard deviation of 19.5. Taking into account the higher degree of chaos of the Lorenz attractor, and given the same problems related to noise, it is hard to expect a better performance of the reservoir computer at following the target trajectory. Figure 4.17 depicts the evolution of the MSE during the autonomous run. The error curve was averaged over 100-timestep intervals. The initial dip corresponds to the teacher-forcing of the reservoir computer with the target signal for 128 timesteps, as discussed in Sect. 4.4. The error exceeds the $10^{-3}$ threshold around the $n = 250$ mark and reaches a constant value of approximately $1.5 \times 10^{-2}$ after less than 1000 timesteps. At this point, the reservoir computer has lost the target trajectory, but keeps on generating a time series with properties similar to the Lorenz series (see Fig. 4.18 for illustration).

Similar to the Mackey-Glass task, we performed a visual inspection of the generated Lorenz series after a long run, and compared the frequency spectra. Figure 4.18

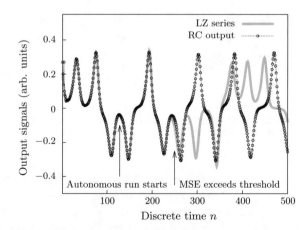

**Fig. 4.16** Example of reservoir computer output signal $y(n)$ (dotted black line) during autonomous runs on the Lorenz task. The system was driven by the target signal (solid grey line) for 128 timesteps before running autonomously (see Sect. 4.4). The MSE threshold was set to $10^{-3}$. The photonic system with $N = 600$ generated 122 correct values in this example, and predicted two switches of the trajectory from one lobe of the attractor to the other

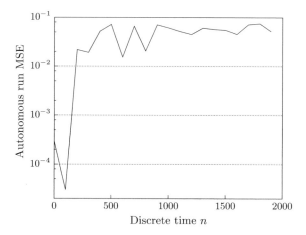

**Fig. 4.17** Evolution of MSE during experimental autonomous generation of the Lorenz chaotic time series (same run as in Fig. 4.16). The error curve, averaged over 100 timesteps, crosses the $10^{-3}$ threshold near $n = 250$. The initial dip corresponds to the warm-up of the reservoir (see Sect. 4.4)

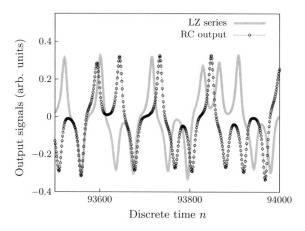

**Fig. 4.18** Output of the experiment (dotted black line) at the end of a long run of 95000 timesteps on the Lorenz task. Although the system does not follow the starting trajectory (solid grey line), it does a fairly good job at emulating the dynamics of the Lorenz system

shows the output of the experiment near the end of a 95000 autonomous run. Although the system is quite far from the target trajectory (plotted in grey) at this point, it is apparent that it has captured the dynamics of the Lorenz system very well. Figure 4.19 displays the Fast Fourier Transforms of the generated time series (dotted black line) and the computed Lorenz series (solid grey line). Unlike the Mackey-Glass system, these frequency spectra do not have any dominant frequencies. That is, the power distribution does not contain any strong peaks, that could have been used as reference points for comparison. Therefore, comparing the two spectra is much more

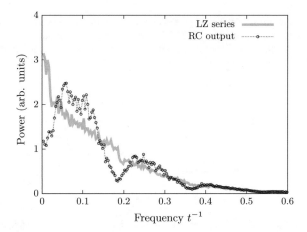

**Fig. 4.19** Comparison of Fast Fourier Transforms of the Lorenz series (solid grey line) and the time series generated by the photonic reservoir computer (dotted black line) during 95000 timesteps. Both spectra are normalised so as to have equal total power. The curves are smoothened by averaging over 50 samples and the plot is limited to lower frequencies (the higher ones being close to zero). Despite some mismatch, the shape of the dotted curve is roughly similar to the grey line

subjective in this case. Although the curves do not match, one can still see a certain similitude between them.

In addition to those visual comparisons, we performed a specific randomness test of the generated series. We exploited an interesting property of the Lorenz dynamics. Since it basically switches between two regions (the wings of the butterfly-like Lorenz attractor), with random transitions from one to the other, one can assign binary "0" and "1" to these regions and thus transform the Lorenz series into a sequence of random bits. We used this technique to check the randomness of the generated series. To this end, we both solved the Lorenz equation and ran the experiment for 95000 timesteps, and converted the resulting time series into two sequences of approximately 2400 bits. The two were then analysed with the ENT program [33]—a well known software for testing random number sequences—with the results shown in Table 4.5. Their interpretation requires a few explanations.

- The first test computes the entropy per byte (8 bits). Since the entropy can be seen as a measure of disorder or randomness, a totally random sequence should have 8 bits of entropy per byte. Both sequences are close to the maximum value, with the Lorenz series being slightly more random.
- The compression, i.e. how efficiently a sequence of bytes could be reduced in size by a compression algorithm, such as e.g. Lempel-Ziv-Renau algorithm, used by the Zip program, is a commonly used indirect method of estimating the randomness of bytes in a file. These algorithms basically look for large repeating blocks, that should not appear in a totally random sequence. Again, both sequences could only be slightly compressed.

**Table 4.5** Results returned by the ENT program for the bit sequences generated by the experiment (RC) and the integrated Lorenz system. The Lorenz sequence shows better figures, but the RC output is not far behind. All these figures are poor compared to common random series, but this is due to the very short sequences used here (roughly 300 bytes)

|                    | RC     | Lorenz |
|--------------------|--------|--------|
| Entropy (/byte)    | 6.6    | 7.1    |
| Compression (%)    | 17     | 10     |
| Mean (byte)        | 134.3  | 125.8  |
| $\pi$              | 2.88   | 3.00   |
| Correlation        | −0.08  | −0.02  |

- The mean value is the arithmetic mean of the data bytes. A random sequence should be evenly distributed around the mean value of 127.5. The Lorenz series is very close to this value, and the RC sequence is fairly close.
- The Monte Carlo method of computing the value of $\pi$ randomly places points inside a square and computes the ratio of points located inside an inscribed circle, that is proportional to $\pi$. This complex test requires a long sequence of bytes to yield accurate results. We note that, nevertheless, both sequences produce a plausible estimation of $\pi$.
- Finally, the serial correlation, i.e. the degree of similarity between the sequence and a delayed copy of itself, in a totally random sequence is zero. Both series present a very low correlation, yet again the Lorenz series demonstrating a better score.

These results do not strictly prove that the generated sequence is random. One obviously has to use a much longer sequence of bits for that task, and should also consider more sophisticated and complete tests, such as Diehard [34] or NIST Statistical Test Suite [35]. The purpose of these tests was to show that the output of the RC generator does not consist of trivial oscillations, that only remotely resemble the Lorenz system. The figures in Table 4.5 show that the randomness of the RC output is similar to the Lorenz system, which gives reasons to believe in the similarity between the properties of the two time series. This, in turn, indicates, that our photonic reservoir computer was capable of learning to effectively emulate the dynamics of the Lorenz chaotic system.

## 4.8  Challenges and Solutions

When I started this project, my FPGA development skills were on a much higher level than during the first project (see Chap. 2). I had a solid grasp on timing closure and a clear idea of what could and what could not be implemented in VHDL. Moreover, several chunks of code could be reused from the previous projects, such as the

Ethernet connection modules and the FMC151 daughtercard interface. But this time the FPGA had to be connected very tightly to the experiment, making the synchronisation of all the module a crucial aspect. In fact, not only the reservoir states had to be sampled at the right moment, but the output signal had to be computed and generated in a very tight timeframe. And since there is no easy way of simulating the opto-electronic reservoir in Xilinx software, the development of the design was mostly done in hardware, which was very time-consuming.

Sooner or later, the FPGA design was operational and the first good results started to come. Sine waves were generated very nicely, and random patterns were stable for hours and even days. The maximum pattern length was significantly lower than expected, though, but I did not pay much attention to this detail at first—I was happy that the experiment generated something stable at all. We then moved on to the Mackey-Glass task and this is where the problems began. I managed to record several very good prediction lengths from the experiment, but the results were very inconsistent. We know now that this is due to the experimental noise, but we did not have that luxury at the time, and I passed a lot of time analysing the setup, trying to figure out what was wrong.

In my previous experiments, unstable outcomes indicated timing problems with the FPGA chip. I double- (even triple-) checked the design and its behaviour only to find that it was functioning flawlessly. Refusing to give up, I developed a very precise numerical model of the opto-electronic reservoir, taking into account all known characteristics of the hardware, up to the high-pass filter—but again, everything except the noise.

The ultimate revelation came when I was comparing individual reservoir states from two different runs and noticed a significant difference between the values. A statistical analysis of ten identical runs of the experiment then showed even higher deviations. Only then I though of adding these deviations to the simulation model and—at last—the inconsistent outcomes of the experiment could be reproduced numerically.

Once we acknowledged the noise as an inevitable "feature" of the experiment, a new question arose—how to evaluate the RC performance on chaotic signals, since it was very bad at following a given trajectory? From observations of the output signal, we could see that it is far from being trivial—but how to show this quantitatively? Since these interrogations were new to our team, we tried to improvise a few simple solutions. The most interesting, in my opinion, was the ENT test for the Lorenz attractor. Far from being a solid proof, it gave us some valuable numbers to quantify the properties of the RC output. I wish we could find something similar for the Mackey-Glass series, though …

## 4.9   Conclusion

The present work demonstrates the potential of output feedback in hardware reservoir computing and constitutes an important step towards autono-mous photonic implementations of recurrent neural networks. Specifically, we presented a photonic reservoir computer capable of generating both sine waves of different frequencies and short random patterns with substantial stability. Moreover, it could emulate the Mackey-Glass time series with a considerably similar frequency spectrum, and a fairly close highest Lyapunov exponent. Finally, it could efficiently capture the dynamics of the Lorenz system and generate a sequence of bytes with similar randomness properties. To the best of our knowledge, this is the first report of these task being implemented on an experimental reservoir computer.

The readout layer of the reservoir computer is processed in real time on a fast FPGA chip. This results in a digital output layer with an analogue reservoir, that nevertheless allows one to investigate many of the issues that will affect a fullyanalogue system. The latter is a much more complicated experiment. Indeed the only analogue output layers implemented so far on experimental reservoir computers were reported in [1–3]. Using them for output feedback would require adding an extra electronic circuit performing the sample and hold, amplification, and multiplication by the input mask. The present experiment allowed us to investigate the novelty that output feedback has to offer to experimental reservoir computing, while anticipating the difficulties and limitations that will affect a fully analogue implementation. Such a two-step procedure, in which one starts with a semi-analogue, semi-digital experiment, is natural, and parallels the development of experimental reservoir computers in which some of the first experiments were only partially analogue, see e.g. [21, 36].

This work allowed to highlight a critical limitation of the present opto-electronic setup, namely the relatively high level of noise generated by the components. While this was not a concern in previous experiments without output feedback, such as [2, 22], it becomes detrimental in this study. Since this noise cannot be averaged out, it propagates back into the system with output feedback and considerably deteriorates the reservoir states. This problem does not seem to have a simple solution. One could rethink the entire experimental setup and rebuild it with new, less noisy components. Switching to a different experimental system, such as the low-noise passive cavity reported in [32], is an alternative approach. In any case, it will probably be difficult to increase the SNR above 60 dB. There may also be algorithmic solutions, such as using conceptors [37, 38].

The high level of experimental noise quickly pushed the reservoir computer, trained to emulate a chaotic system, away from a given trajectory. This outcome initiated the search for alternative methods for the evaluation of the experiment performance. In this work, we introduced a few simple techniques, based on standard signal analysis methods, such as statistics of the prediction length and visual comparison of the time series and their frequency spectra after a long autonomous run. We have also proposed case-specific methods, such as the randomness test, that could only be applied to the Lorenz time series. Overall, these are the first steps towards the answer to a very general question: given a noisy emulator of a known chaotic system, how best to evaluate its performance? It will be interesting to understand the relationship between the performance obtained on the estimators above, and the properties of the chaotic system, such as the Lyapunov exponents or the dimension and geometry of the chaotic attractor. These questions should lead to a rich new direction of enquiry in the theory of nonlinear dynamics and complex systems.

The addition of output feedback allows experimental reservoir computers to solve much more complex tasks than without output feedback. Future work could address nonlinear computations that depend on past information and that require persistent memory [39], FORCE training [40] (however, this algorithm requires that the learning time scale be short compared to the reservoir time scale) and applications such as frequency modulation [15], or implementation of conceptors [37, 38]. Ideally, a fully analogue feedback should be implemented, like in e.g. [41] (see also Chap. 5), rather than the digital feedback demonstrated here. Therefore, the present work is a first step towards realising these additional applications.

Finally, going back to the question of biological implementation, our work shows that the biologically plausible structure of reservoir computing [18–20] can be trained to generate highly complex temporal patterns, both periodic and chaotic, even in the presence of moderate levels of noise. Whether nature in fact implements this mechanism remains to be seen, and will depend, amongst other aspects, on the amount of noise present in biological implementations of reservoir computing, and whether there exist biologically plausible training mechanisms for this kind of signal generation.

# References

1. Smerieri, Anteo, François Duport, Yvan Paquot, Benjamin Schrauwen, Marc Haelterman, and Serge Massar. 2012. Analog readout for opticalreservoir computers. In *Advances in neural information processing systems*, 944–952.
2. Duport, François, Anteo Smerieri, Akram Akrout, Marc Haelterman, and Serge Massar. 2016. Fully analogue photonic reservoir computer. *Scientific Report* 6: 22381.
3. Vinckier, Quentin, Arno Bouwens, Marc Haelterman, and Serge Massar. 2016. Autonomous all-photonic processor based on reservoir computingparadigm. In *Conference on lasers and electro-optics*. Optical Societyof America. SF1F.1.
4. Antonik, Piotr, Marc Haelterman, and Serge Massar. 2017. Brain-inspiredphotonic signal processor for generating periodic patterns and emulatingchaotic systems. *Physical Review Applied* 7: 054014.
5. Antonik, Piotr, Michiel Hermans, Marc Haelterman, and Serge Massar. 2017. Random pattern and frequency generation using a photonic reservoircomputer with output feedback. *Neural Processing Letters*: 1–14.
6. Zhang, G. Peter. 2012. Neural networks for time-series forecasting. In *Handbook of natural computing*, ed. Grzegorz Rozenberg, ThomasBack, and Joost N. Kok, 461–477. Berlin: Springer.
7. Wyffels, Francis and Benjamin Schrauwen. 2010. A comparative study of Reservoir Computing strategies for monthly time series prediction. *Neurocomputing* 73 (10–12): 1958–1964.
8. Antonik, Piotr, Michiel Hermans, François Duport, Marc Haelterman, and Serge Massar. 2016. Towards pattern generation and chaotic series predictionwith photonic reservoir computers. In *SPIE's 2016 laser technology and industrial laser conference*, vol. 9732, 97320B.
9. Xu, Meiling, Min Han, and Shunshoku Kanae. 2016. L1/2 norm regularizedecho state network for chaotic time series prediction. In *APNNS's 23th international conference on neural information processing (ICONIP)*, vol. 9886. LNCS, 12–19.
10. *The 2006/07 forecasting competition for neural networks and computational intelligence.* http://www.neural-forecasting-competition.com/NN3/.

11. Jaeger, Herbert, and Harald Haas. 2004. Harnessing nonlinearity: Predictingchaotic systems and saving energy in wireless communication. *Science* 304: 78–80.
12. Wyffels, Francis, Benjamin Schrauwen, and Dirk Stroobandt. 2008. Stableoutput feedback in reservoir computing using ridge regression. In *International conference on artificial neural networks*, 808–817. Berlin: Springer.
13. Caluwaerts, Ken, Michiel D'Haene, David Verstraeten, and Benjamin Schrauwen. 2013. Locomotion without a brain: Physical reservoir computing in tensegrity structures. *Artificial Life* 19 (1): 35–66.
14. Reinhart, Rene Felix, and Jochen Jakob Steil. 2012. Regularization and stabilityin reservoir networks with output feedback. *Neurocomputing* 90: 96–105.
15. Wyffels, Francis, Jiwen Li, Tim Waegeman, Benjamin Schrauwen, and Herbert Jaeger., 2014. Frequency modulation of large oscillatory neural networks. *Biological Cybernetics* 108 (2): 145–157.
16. Antonik, Piotr, Michiel Hermans, Marc Haelterman, and Serge Massar. 2016. Towards adjustable signal generation with photonic reservoir computers. In *25th international conference on artificial neural networks*, vol. 9886.
17. Jaeger, Herbert. 2007. Echo state network. *Scholarpedia* 2 (9): 2330.
18. Maass, Wolfgang, Thomas Natschlager, and Henry Markram. 2002. Realtimecomputing without stable states: A new framework for neural computation based on perturbations. *Neural computation* 14: 2531–2560.
19. Yamazaki, Tadashi, and Shigeru Tanaka. 2007. The cerebellum as a liquid state machine. *Neural Networks* 20 (3): 290–297.
20. Rossert, Christian, Paul Dean, and John Porrill. 2015. At the edge of chaos: How cerebellar granular layer network dynamics can provide the basis for temporal filters. *PLOS Computational Biology* 11 (10): 1–28. Oct.
21. Appeltant, Lennert, Miguel Cornelles Soriano, Guy Van der Sande, Serge Massar, JanDanckaert, Joni Dambre, Benjamin Schrauwen, Claudio R. Mirasso, and Ingo Fischer. 2011. Information processing using a single dynamical node as complex system. *Nature Communications* 2: 468.
22. Paquot, Yvan, François Duport, Anteo Smerieri, Joni Dambre, Marc Haelterman Benjamin-Schrauwen, and Serge Massar. 2012. Optoelectronic reservoir computing. *Scientific Reports* 2: 287.
23. Larger, Laurent, M.C. Soriano, Daniel Brunner, L Appeltant, Jose M Gutierrez, Luis Pesquera, Claudio R Mirasso, and Ingo Fischer. 2012. Photonicinformation processing beyond Turing: An optoelectronic implementation of reservoir computing. *Optics Express* 20: 3241–3249.
24. Mackey, Michael C., and Leon Glass. 1977. Oscillation and chaos in physiologicalcontrol systems. *Science* 197 (4300): 287–289.
25. Lorenz, Edward N. 1963. Deterministic nonperiodic flow. *Journal of the atmospheric sciences* 20 (2): 130–141.
26. Antonik, Piotr, François Duport, Michiel Hermans, Anteo Smerieri, Marc Haelterman, and Serge Massar. 2016. Online training of an Opto- electronic reservoir computer applied to Real-Time channel equalization. *IEEE Transactions on Neural Networks and Learning Systems* 28 (11): 2686–2698.
27. Farmer, Doyne J. 1982. Chaotic attractors of an infinite-dimensional dynamicalsystem. *Physica D: Nonlinear Phenomena* 4 (3): 366–393.
28. Atkinson, Kendall E. 2008. *An introduction to numerical analysis*. Wiley.
29. Hirsch, Morris W., Stephen Smale, and Robert L. Devaney. 2003. *Differential equations, dynamical systems, and an introduction to chaos*. Boston: Academic press.
30. Horowitz, Paul, and Winfield Hill. 1980. *1980*. The art of electronics: Cambridge University Press.
31. Oppenheim, A.V., and R.W. Schafer. 1989. *Discrete-time signal processing*. Prentice-Hall signal processing series: Prentice Hall. ISBN 9780132162920. https://books.google.fr/books?id=bPhSAAAAMAAJ.

32. Vinckier, Quentin, François Duport, Anteo Smerieri, Kristof Vandoorne, Peter Bienstman, Marc Haelterman, and Serge Massar. 2015. High-performancephotonic reservoir computer based on a coherently driven passive cavity. *Optica* 2 (5): 438–446.
33. Walker, John. *ENT program*. http://www.fourmilab.ch/random/.
34. Marsaglia, George. *The Marsaglia random number CDROM including the Diehard Battery of Tests of Randomness*. http://stat.fsu.edu/pub/diehard/.
35. Rukhin, Andrew, Juan Soto, James Nechvatal, Miles Smid, and Elaine Barker. 2001. *A statistical test suite for random and pseudorandom number generators for cryptographic applications*. National Institute of Standards and Technology: Technical report.
36. Martinenghi, Romain, Sergei Rybalko, Maxime Jacquot, Yanne Kouomou Chembo, and Laurent Larger. 2012. Photonic nonlinear transient computingwith multiple-delay wavelength dynamics. *Physical Review Letters* 108: 244101.
37. Jaeger, Herbert. 2014. Conceptors: An easy introduction. In *CoRR abs/1406.2671*.
38. Jaeger, Herbert. 2014. Controlling recurrent neural networks by conceptors. In *CoRR abs/1403.3369*.
39. Kovac, André David, Maximilian Koall, Gordon Pipa, and Hazem Toutounji. 2016. Persistent memory in single node delay-coupled reservoir computing. *PLOS ONE* 11 (10): 1–15.
40. Sussillo, David, and L.F. Abbott. 2009. Generating coherent patterns ofactivity from chaotic neural networks. *Neuron* 63 (4): 544–557.
41. Antonik, Piotr, Marc Haelterman, and Serge Massar. 2017. Online trainingfor high-performance analogue readout layers in photonic reservoir computers. *Cognitive Computation* 9: 297–306.

# Chapter 5
# Towards Online-Trained Analogue Readout Layer

This chapter presents the last project that I started with the OPERA-Photonique group (the next and last Chap. 6 will cover my internship at the University of Texas at Austin, USA). At the moment of writing these lines, we published some interesting results, but they are based on numerical simulations only.

Ironically, building an improved analogue readout layer should have been the corner stone of my thesis: this was stated in my funding application, and this probably was the main reason why the lab purchased a FPGA in the first place.

How comes that this project was not completed? I can see two explanations. First, we found an easier way to achieve the ultimate goal. In fact, one of the key motivations for an efficient analogue readout was the possibility of feeding the output signal back into the reservoir. Chapter 4 shows how this can be done with a much more simple digital layer, and what new features can be obtained out of it. The second reason is a purely personal one. I had the aspiration to explore new research environments and when such an occasion appeared, I gladly took it. But I am getting ahead of myself—more on that in the next chapter (Chap. 6).

In fine, I believe we replaced my original thesis project with an even better plan. Instead of investing time into the development of an analogue readout layer—a direction than has already been explored to some degree [1–3], we focused on the in-depth study of an experimental reservoir computer with output feedback (although digital)—a truly novel direction, that nobody has investigated so far. But the story of the analogue readout layer is not over yet. Although the present thesis will only contain numerical results of this work, I still have a couple of month in front of me to do more.

This chapter, based on our paper [4], is much shorter than the previous ones: it does not contain sections on the FPGA design or on challenges and solutions, since it does not present an actual physical experiment.

© Springer International Publishing AG, part of Springer Nature 2018
P. Antonik, *Application of FPGA to Real-Time Machine Learning*,
Springer Theses, https://doi.org/10.1007/978-3-319-91053-6_5

## 5.1   Introduction

The major drawback in experimental implementations of reservoir computing, listed in Sect. 1.2, is the absence of efficient readout mechanisms: the states of the neurons are collected and post-processed on a computer, severely reducing the processing speeds and thus limiting the applicability. An analogue readout would resolve this issue, as suggested in [5]. This research direction has already been investigated experimentally in [1–3], but all these implementations suffered from significant performance degradation due to the complex structure of the readout layer. Indeed the approach used in these works was to characterise with high accuracy the linear output layer, whereupon it was possible to compute offline the output weights. However it is virtually impossible to characterise each hardware component of the setup with sufficient level of accuracy. Furthermore the components in the output layer may have a slight nonlinear behaviour. It follows that this approach does not work satisfactorily, as is apparent from the performance degradation reported in [2].

In this work we address the above issues with the same online learning approach we discussed in Chap. 2. Online training has attracted much attention in the machine learning community because it allows to optimise the system gradually, as the input data becomes available. As we have shown in Chap. 2, it can also easily cope with non-stationary input signal, whose characteristics change with time, as the online approach can keep the model updated according to variations in the input. Finally, in the case of hardware systems, online training can easily cope with drifts in the hardware, as the system will adapt to gradual changes in the hardware components [6, 7].

In the context of reservoir computing, the online training implements a gradient descent: it gradually changes the output layer to adapt to the task. More precisely the output layer is characterised by a series of parameters (the readout weights), and in online training these weights are adjusted in small increments, so that the output of the system gets closer to the target signal. The important point in the present context is that, compared to previously used offline methods, in online training based on gradient descent no assumption is necessary about how these weights contribute to the output signal. That is, it is not necessary to model the output layer. Furthermore, the transfer function of the readout layer could in principle be nonlinear. Here we show, using realistic numerical simulations, how these features could be highly advantageous for training hardware reservoir computers.

For concreteness, we consider in simulations the "same old" opto-electronic reservoir computing setup, introduced in Sect. 1.2.4. We add to this setup an analogue layer that is now trained online by an FPGA chip processing the simple gradient descent algorithm in real time, as in [8] and Chap. 2. The readout layer consists of a simple Resistor-Capacitor (RC) circuit (as in [1]), instead of a more complicated RLC circuit (consisting of a resistor $R$, an inductor $L$ and a capacitor $C$) that was used to increase the amplitude of the output signal in [2].

We investigate the performance of this setup through numerical simulations on two benchmark tasks and show that previously encountered difficulties are almost entirely alleviated by the online training approach. In other words, with a relatively simple analogue readout layer, trained online, and without any modelling of the underlying processes, we obtain results similar to those produced by a digital layer, trained offline. We also explore a special case with a nonlinear readout function and show that this complication does not decrease much the performance of the system. This work thus brings an interesting solution to an important problem in the hardware reservoir computing field.

## 5.2 Methods

This section occupies almost two pages in the original paper [4], but seems completely useless in this thesis. That is, all methods used in this work have already been introduced. The reservoir computing principles have been widely discussed in Sect. 1.1.3. The gradient descent algorithm has been outlined in Sect. 2.3.1. And the two benchmark tasks used here—wireless channel equalisation and NARMA10—have been introduced in Sect. 1.1.4. Therefore, without further ado, let us move straight to the proposed experimental setup, which eventually contains something new—an analogue readout layer.

## 5.3 Proposed Experimental Setup

Figure 5.1 depicts the proposed experimental setup that we have investigated using numerical simulations. The opto-electronic reservoir needs no introduction (see Sect. 1.2.4). The analogue readout layer, however, is a different story.

### 5.3.1 Analogue Readout Layer

The analogue readout layer uses the same scheme as proposed in [1]. The optical power it receives from the reservoir is split in two. Half is sent to the readout photodiode (TTI TIA-525I), and the resulting voltage signal, containing the reservoir states $x_i(n)$, is recorded by the FPGA for the training process (see Eq. 2.4). The other half is modulated by a dual-output Mach-Zehnder modulator (EOSPACE AX-2X2-0MSS-12) which applies the readout weights $w_i$, generated by the DAC of the FPGA. The outputs of the modulator are connected to a balanced photodiode (TTI TIA-527), which produces a voltage level proportional to difference of the light intensities received at its two inputs. This allows to multiply the reservoir states by both positive and negative weights [1]. The summation of the weighted states is performed by a low-pass RC filter. The resistance R of the filter, not shown on the scheme, is the 50 Ω output impedance of the balanced photodiode. The resulting

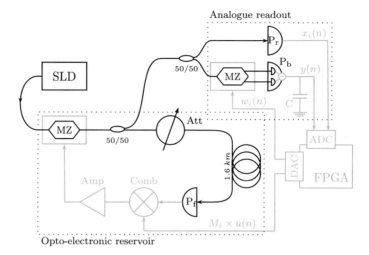

**Fig. 5.1** Scheme of the proposed experimental setup. The optical and electronic components are shown in black and grey, respectively. The reservoir layer consists of an incoherent light source (SLD), a Mach-Zehnder intensity modulator (MZ), a 50/50 beam splitter, an optical attenuator (Att), an approximately 1.6 km fibre spool, a feedback photodiode (P$_f$), a resistive combiner (Comb) and an amplifier (Amp). The analogue readout layer contains another 50/50 beam splitter, a readout photodiode (P$_r$), a dual-output intensity modulator (MZ), a balanced photodiode (P$_b$) and a capacitor (C). The FPGA board generates the inputs and the readout weights, samples the reservoir states and the output signal, and trains the system

output signal, proportional to $y(n)$, is also recorded by the FPGA, for training and performance evaluation.

Let us compute explicitly the output of the analogue readout layer. The capacitor integrates the output of the balanced photodiode with an exponential kernel and a time constant $\tau$. The impulse response of the RC filter is given in [9]

$$h(t) = \frac{1}{RC}e^{\frac{-t}{RC}} = \frac{1}{\tau}e^{\frac{-t}{\tau}}, \tag{5.1}$$

the voltage $Q(t)$ on the capacitor is then given by

$$Q(t) = \int_{-\infty}^{t} X(s)W(s)h(t-s)ds, \tag{5.2}$$

where $X(t)$ is the continuous signal, containing the reservoir states, and $W(t)$ are the readout weights, applied to the dual-output intensity modulator. The output $y(n)$ is given by the charge on the capacitor at the discrete times $t = nT$:

$$y(n) = Q(nT). \tag{5.3}$$

Since $X(t)$ and $W(t)$ are piecewise functions $X(t) = x_i(n)$ and $W(t) = w_i$ for $t \in [\theta(i-1), \theta i]$, where $\theta = T/N$ is the duration of one neuron, we can approximate the integration by a discrete summation to obtain

$$
\begin{aligned}
y(n) &= \theta \sum_{i=1}^{N} w_i \left( \sum_{k=0}^{\infty} x_i(n-k) h(N-i-Nk) \right) \\
&= \frac{\theta}{\tau} \sum_{i=1}^{N} w_i \left( \sum_{k=0}^{\infty} x_i(n-k) e^{-\rho(N-i-Nk)} \right),
\end{aligned}
\tag{5.4}
$$

where we have introduced the RC integrator ratio $\rho = \theta/\tau$.

The readout layer output $y(t) = Q(t)$ is thus a linear combination of the reservoir states $x_i$, weighted by $w_i$ and by the exponential kernel of the RC filter. Note that contrary to usual reservoir computer outputs (see e.g. Eq. 1.6), in Eq. 5.4 the output at time $n$ depends not only on the current states $x_i(n)$, but also on the states at previous times $x_i(n-k)$.

In the previous experimental investigation of the analogue readout [2], the readout weights $w_i$ were computed using ridge regression [10] (see also Sect. 1.1.3), assuming an output signal given by Eq. 1.6. But since the experiment produced an output similar to Eq. 5.4 instead, the readout weights needed to be corrected appropriately. For more details, we refer to the original paper [2]. In the present work, the weights $w_i$ are adjusted gradually to match the reservoir output signal $y(n)$ with the target output $d(n)$ (see Sect. 2.3.1), without any assumptions about how these weights actually contribute to the output signal $y(n)$. This is a much easier tasks, which allows to obtain better experimental results, as will be shown in Sect. 5.5.

## 5.3.2 FPGA Board

Similar to my previous experiments, the reservoir computer is operated by a FPGA chip. We envision using the same Xilinx ML605 evaluation board, paired with the 4DSP FMC151 daughter card. For this experiment, we will need both ADCs and both DACs.

The FPGA generates the input signal $M_i \times u(n)$ and sends it into the opto-electronic reservoir. After recording the resulting reservoir states $x_i(n)$ from one delay loop, it executes the simple gradient descent algorithm in order to update the readout weights $w_i(n+1)$. These are sent to the readout layer and used to generate the output signal $y(n)$, also recorded by the FPGA.

## 5.4   Numerical Simulations

All numerical experiments were performed in Matlab. We used a custom model of our reservoir computer, based on previous investigations [8, 11, 12] (see also Sects. 2.6.3 and 4.6), that has been shown to emulate very well the dynamics of the real system. The simulations were performed in discrete time, and took into account the internal structure of the reservoir computer described above, such as the ring-like topology, the sine nonlinearity and the analogue readout layer with an RC filter. The simulations allow to try out different configurations and to scan various experimental parameters, including values that are impossible to achieve experimentally or imposed by the hardware. All simulations were performed on a dedicated high-performance workstation with 12-core CPU and 64 Gb RAM. Since the convergence of the gradient descent algorithm is quite slow, we limited our investigations to a fast update rate $k = 10$ (see Eq. 2.5), so that each simulation lasted about 24 h.

   The principal goal of the simulations was to check how the online learning approach would cope with experimental difficulties encountered in previous works [1, 2]. To that end, we gathered a list of possible issues and scanned the corresponding experimental parameters in order to check the system performance. In particular, we investigated the following parameters:

- The RC integrator ratio $\rho$. This is the most important parameter of the analogue readout layer. While its accurate measure is not required in our setup—since we do not correct the readout weights $w_i$—it defines the integration span of the filter, and thus the reservoir states that contribute to the output signal. It can thus significantly influence the results. Another question of importance is how dependent the system performance is on the exact value of $\rho$.
- The MZ modulator bias. Mach-Zehnder modulators need to be applied a constant voltage to maintain their transfer function symmetric around zero. The devices we were using up to now are subject to slight drifts over time, often resulting in a non-perfectly symmetric response. We thus checked in simulations whether such an offset would impact the results.
- The DAC resolution. The precision of the DACs on the FMC151 daughtercard is limited to 16 bits. Numerical investigations have shown that the precision of readout weights has a significant impact on the performance, see e.g. [13–15]. We thus checked whether the resolution available is enough for this experiment.

Besides these potentially problematic parameters, we also scanned the input and feedback gain parameters (denoted by $\beta$ and $\alpha$ in Eq. 1.5) in order to find the optimal dynamics of the reservoir for each task.

   In a separate set of simulations, we investigated the applicability of the proposed method to nonlinear readout layers. That is, we checked whether the simple gradient descent method would still work with a nonlinear response of the analogue readout layer with respect to the reservoir states $x_i(n)$ (see Eq. 5.4). We picked two "saturation" functions of sigmoid shape. This choice arises from the transfer function of common light detectors that are linear at lower intensities and become nonlinear at higher intensities. We used the following functions: a logistic function, given by

$$g_{lg}(x) = \frac{2}{1 + e^{-2x}} - 1, \tag{5.5}$$

and the hyperbolic tangent function, given by

$$g_{ht}(x) = 0.6 \tanh(1.8x). \tag{5.6}$$

These functions, $g_{lg}$ and $g_{ht}$, do not model any particular photodiode, but are two simple examples that allow us to address the above question. Both functions are plotted in Fig. 5.4, together with a linear response, for comparison.

We investigated two possible nonlinearities in the output layer. In the first case, the readout photodiode ($P_r$ in Fig. 5.1) produces a nonlinear response, while the balanced photodiode ($P_b$ in Fig. 5.1) remains linear. This scenario, that we shall refer to as "nonlinear readout", allows one to investigate what happens when the reservoir states $x_i$ used to compute the output signal $y(n)$ (see Eq. 1.6) differ from those employed to update the readout weights (see Eq. 2.4). Thus, in this case the update rule (Eq. 2.4) for the output weights becomes

$$w_i(n + 1) = w_i(n) + \lambda (d(n) - y(n)) g(x_i(n)), \tag{5.7}$$

where $g$ is given by either Eq. 5.5 or Eq. 5.6, while the output layer is given by Eq. 5.4.

In the second case, called "nonlinear output", the readout photodiode is linear, but the balanced photodiode exhibits a saturable behaviour. In this case the update rule Eq. 2.4 for the output weights is unchanged, but the output layer Eq. 5.4 becomes

$$y(n) = \frac{\theta}{\tau} \sum_{i=1}^{N} w_i \left( \sum_{k=0}^{\infty} g(x_i(n - k)) e^{-\rho(N-i-Nk)} \right). \tag{5.8}$$

Note that we have only considered cases with just one nonlinear photodiode, so as to check whether the difference between the reservoir states used for training and those to compute the readout (see Eqs. 2.4 and 1.6, respectively) would degrade the performance of the system. The scenario with both nonlinear photodiodes is hence more simple, as the reservoir states are the same in both equations. One could consider the case with two photodiodes exhibiting different nonlinear behaviours. In that situation, similar to the results we will show in Sect. 5.5, we expect the algorithm to cope with the difference up to a certain point, before running into troubles. For this reason, we leave that scenario for future investigations.

## 5.5 Results

For clarity, the results are split in two sections. First, we discuss the influence of the key parameters, listed in Sect. 5.4, in the case of a linear readout layer. Then, we consider the two nonlinear scenarios described above.

### 5.5.1 Linear Readout: RC Circuit

For each of the two tasks considered here, we performed three kinds of simulations: we scanned the RC integrator ratio $\rho = \theta/\tau$ in the first simulation, the MZ bias in the second, and the resolution of the DAC in the third. Furthermore, since different values of these parameters may work better with different dynamics of the reservoir, we also scanned the input gain $\beta$ and the feedback gain $\alpha$ in all three simulations independently, and applied the optimal values in each case.

For both tasks, we used a network with $N = 50$ neurons, as in most previous experimental works [2, 11, 16, 83]. The reservoir was trained on 83000 inputs, with an update rate $k = 10$, and then tested over $10^5$ symbols for the channel equalisation task and $10^4$ inputs for NARMA10 task. For statistical purposes, we ran each simulation 10 times, with different random input masks. In the following figures, averaged results over the masks are plotted, while the error bars give the standard deviation over the different input masks. Results related to the channel equalisation task are plotted with solid lines, while dashed lines correspond to those for NARMA10.

For the channel equalisation task, our system yields SERs between $10^{-4}$ and $10^{-3}$ depending on the input mask, as summarised in Table 5.1 (first line). This is comparable to previous experiments with the same opto-electronic reservoir: error rates of order of $10^{-4}$ were reported in [11] using a digital readout and in [2] with an analogue readout, using an RLC filter. The first experimental analogue system, using a simple RC circuit, as we did in this work, performed significantly worse, with SER of order of $10^{-2}$ [1]. That is, online learning does not outperform other methods, but allows to obtain significantly better results with a simpler setup.

As for the NARMA10 task, we obtain a NMSE of $0.20 \pm 0.02$. Previous experiments with a digital readout layer produced $0.168 \pm 0.015$ [11] and $0.107 \pm 0.012$ [17]. With an analogue readout layer, the best NMSE reported was $0.230 \pm 0.023$ [2]. Our system thus slightly outperforms the analogue approach, and gets close to the digital one, except for the very good result obtained with a different reservoir, based on a passive cavity [17]. Again, our results were obtained with a simple setup and no modelling of the readout, contrary to [2].

Furthermore, the error rates obtained here can be significantly lowered with more training, as has been demonstrated numerically and experimentally in [8] (see also Sect. 2.6.2). To keep reasonable simulation times (about 24 h per simulation), we limited the training to 83000 input values, with an update rate $k = 10$. Higher update rates can be used experimentally, since running the opto-electronic setup is much

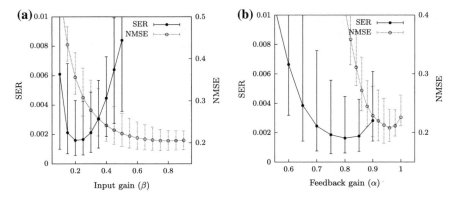

**Fig. 5.2** Reservoir computer performances for different input ($\beta$) and feedback ($\alpha$) gains (solid lines: channel equalisation, dashed lines: NARMA10). **a** While channel equalisation is relatively sensitive to $\beta$, NARMA10 works well in a wide range of values. Note that although it seems that higher input gain would give better results, the dashed curve actually rises slightly for large $\beta$, and the optimum input gain is around 0.8. **b** Both tasks require a system with significant memory (feedback gain at least $\alpha = 0.8$), and even a near-chaotic regime for NARMA10 ($\alpha = 0.95$)

faster than simulating it. We thus expect to obtain lower error rates experimentally with longer training sets and update rates up to $k = 200$. To illustrate this point with results reported in [8] (and discussed in Sect. 2.6.1), short training sets with $k = 10$ yielded SERs of order of $10^{-4}$ for the channel equalisation task. Increasing $k$ up to 200 allowed to decrease the error rate down to $5.7 \times 10^{-6}$.

Figures 5.2a, b show the influence of input and feedback gain parameters on the performance of the system. All curves present a pronounced minimum, except for the input gain $\beta$ for the NARMA10 task, where values above 0.6 seem to produce comparable results. Note that the channel equalisation task requires a low input signal with $\beta = 0.2$, while NARMA10 works best with stronger input and $\beta = 0.8$. As for the feedback gain, NARMA10 shifts the system close to the chaotic regime with $\alpha = 0.95$, while channel equalisation works better with $\alpha = 0.8$.

Figure 5.3a shows the results of the scan of the RC integrator ratio $\rho$. Both tasks work well on a relatively wide range of values, with NARMA10 much less sensitive to $\rho$ than channel equalisation. In particular, the channel is equalised best with $\rho = 0.03$. With $N = 50$, this corresponds to $\tau = T/0.03N = 5.29\,\mu s$, which is shorter than the roundtrip time $T = 7.94\,\mu s$. On the other hand, NARMA10 output is best reproduced with $\rho = 0.003$, which yields $\tau = T/0.003N = 52.93\,\mu s$. This is significantly longer than the roundtrip time $T$, meaning that reservoir states from previous time steps are also taken into account for computation of an output value. This is not surprising, since NARMA10 function has a long memory (see Eq. 1.20). However, this memory effect in the readout layer is not crucial, as the system performs equally well with higher $\rho$ and thus lower $\tau$. All in all, these results are very encouraging for upcoming experiments, as they show that an accurate choice of capacitor is not crucial for the performance of the system.

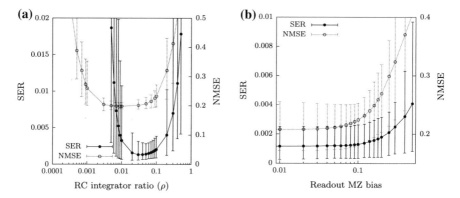

**Fig. 5.3** Impact of the RC integrator ratio ($\rho$) and the readout MZ modulator bias on the reservoir computer performance (solid lines: channel equalisation, dashed lines: NARMA10). **a** Ratios within $\rho \in [0.03, 0.08]$ are suitable for channel equalisation and $\rho \in [0.002, 0.07]$ for NARMA10. Remarkably, inaccurate choice of $\rho$, and thus $\tau$, will not result in significant performance loss, as long as the value lies approximately in the optimal interval. **b** Although the NARMA10 task is more sensitive to this bias, both tasks work reasonably well with a bias up to 0.06, which is superior to expected experimental deviations

Figure 5.3b illustrates the impact of the bias of the readout Mach-Zehnder modulator on the reservoir computer performance. NARMA10 task is clearly more affected by this offset, as the NMSE grows quickly from a bias of roughly 0.06. The SER, on the other hand, stays low until 0.1. For a MZ modulator with $V_\pi = 4.5$ V (see Table 1.1) this corresponds to a tolerance of roughly 0.1 V, which is superior to expected experimental deviations. The Hameg power supply that we use to bias the modulator (see Table 1.1) has a resolution of 0.001 V.

Figure 5.4a shows that the 16-bit DAC resolution is not an issue for this experiment, as the minimal precision required for good performance is 8 bits, for both tasks.

### 5.5.2   Nonlinear Readout

Table 5.1 sums up the results obtained with a nonlinear readout layer. We used optimal experimental parameters, as described above, and generated new sets of data for the training and test phases. We investigated two scenarios and used two functions of sigmoid shape, $x \to g_{\mathrm{lg}}(x)$ and $x \to g_{\mathrm{ht}}(x)$, as described in Sect. 5.4. The system was trained over 83000 inputs, with an update rate $k = 10$, and tested over $10^5$ symbols for the channel equalisation task and $10^4$ inputs for NARMA10. We report error values averaged over 10 trials with different random input masks, as well as the standard deviations. The figures show that the performance deterioration is more manifest with the hyperbolic tangent function $g_{\mathrm{ht}}$, as it is much more nonlinear than

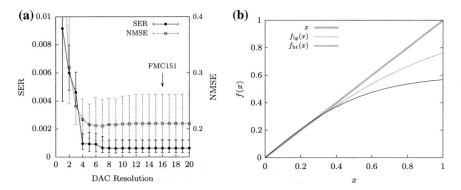

**Fig. 5.4 a** Impact of the DAC resolution on the reservoir computer performance (solid lines: channel equalisation, dashed lines: NARMA10). The results show that the 16-bit resolution of the FMC151 daughtercard is sufficient for this application. **b** Nonlinear response curves of the photodiodes: hyperbolic tangent function $g_{ht}$ (solid line) and logistic function $g_{lg}$ (dotted line). The linear response is plotted with a thick grey line

**Table 5.1** Summary of reservoir computer performances with nonlinear readout layers, measured with error metrics related to the tasks considered here. All values are averaged over 10 random input masks and presented with their standard deviations. We used two functions with sigmoid shape to model the response of the photodiodes. We investigated two scenarios: in the "nonlinear readout" configuration, the readout photodiode $P_r$ is nonlinear, while the balanced photodiode $P_b$ is linear, and vice versa in the "nonlinear output" scheme. The linear case $x \to x$ is shown for comparison. For both tasks, the added nonlinearity does not significantly deteriorate the system performance

| Readout | Transfer function | Chan. Equal. (SER $\times 10^{-3}$) | NARMA10 (NMSE) |
|---|---|---|---|
| Linear | $x$ | $1.1 \pm 0.7$ | $0.20 \pm 0.02$ |
| Nonlinear readout | $g_{lg}(x)$ | $1.3 \pm 0.9$ | $0.21 \pm 0.03$ |
| | $g_{ht}(x)$ | $1.2 \pm 0.8$ | $0.21 \pm 0.02$ |
| Nonlinear output | $g_{lg}(x)$ | $2.0 \pm 1.6$ | $0.21 \pm 0.02$ |
| | $g_{ht}(x)$ | $2.5 \pm 2.1$ | $0.21 \pm 0.01$ |

the logistic function $g_{lg}$. Overall, the added nonlinearity does not have a significant influence on the results in both cases. The SER roughly doubles, at most, for the channel equalisation task. The impact on NARMA10 is barely noticeable, as the error increase of 5% is smaller than the standard deviation. Using offline training on the same system (i.e. with nonlinear output) we observed an increase of the SER by one order of magnitude for the channel equalisation task, and a 30% increase of the NMSE with the NARMA task. These results show that online training is very well suited for experimental analogue layers, as it can cope with realistic components that do not have a perfectly linear response.

## 5.6   Conclusion

In this work we proposed the online learning technique to improve the performance of analogue readout layers for photonic reservoir computers. We demonstrated an opto-electronic setup with an output layer based on a simple RC filter, and tested it, using numerical simulations, on two benchmark tasks. Training the setup online, with a simple gradient descent algorithm, allowed to obtain the same level of performance as with a digital readout layer. Furthermore, our approach does not require any modelling of the underlying hardware, and is robust against possible experimental imperfections, such as inaccurate choice of parameters or components. It is also capable of dealing with a nonlinearity in the readout layer, such as saturable response of the photodiodes. Finally, we expect the conclusions of the present investigation, namely the advantage of online training, to be applicable to all hardware reservoir computers, and not restricted to the delay dynamical opto-electronic systems used for the sake of illustration in the present work.

The results reported in this work will serve as a basis for future investigations involving experimental validation of the proposed method. Experimental realisation of an efficient analogue readout layer would allow building fully-analogue high-performance RCs, abandon the slow digital post-processing and take full advantage of the fast optical components. Such setups could be applied to emerging communication channels [18]. Furthermore, fully-analogue setups would open the possibility of feeding the output signal back into the reservoir, just as we did digitally in [12] (see Chap. 4). Replacing the digital layer with an analogue solution would significantly increase the speed of such generators. Our work thus brings an efficient solution to an important problem in the reservoir computing field, potentially leading to a significant speed gain and a broader range of applications.

## References

1. Smerieri, Anteo, François Duport, Yvan Paquot, Benjamin Schrauwen,Marc Haelterman, and Serge Massar. 2012. Analog readout for optical reservoir computers. In *Advances in neural information processing systems*, pp. 944–952.
2. Duport, François, Anteo Smerieri, Akram Akrout, Marc Haelterman, and Serge Massar. 2016. Fully analogue photonic reservoir computer. *Scientific Reports* 6: 22381.
3. Vinckier, Quentin, Arno Bouwens, Marc Haelterman, and Serge Massar. 2016. Autonomous all-photonic processor based on reservoir computing paradigm. In *Conference on lasers and electro-optics*. Optical society of America, SF1F.1.
4. Antonik, Piotr, Marc Haelterman, and Serge Massar. 2017. Online training for high-performance analogue readout layers in photonic reservoir computers. *Cognitive Computation* 9: 297–306.
5. Woods, Damien, and Thomas J. Naughton. 2012. Optical computing: Photonic neural networks. *Nature Physics* 8 (4): 257–259.
6. Léon, Bottou. 1998. *Online algorithms and stochastic approximations*. In Online learning and neural networks: Cambridge University Press.
7. Shalev-Shwartz, Shai. 2012. Online learning and online convex optimization. *Foundations and Trends in Machine Learning* 4 (2): 107–194.

8. Antonik, Piotr, François Duport, Michiel Hermans, Anteo Smerieri, Marc Haelterman, and Serge Massar. 2017. Online training of an opto-electronic reservoir computer applied to real-time channel equalization. *IEEE Transactions on Neural Networks and Learning Systems* 28 (11): 2686–2698.

9. Horowitz, Paul, and Winfield Hill. 1980. *The art of electronics*. Cambridge University Press.

10. Tikhonov, Andrei Nikolaevich, A.V. Goncharsky, V.V. Stepanov, and Anatoly G. Yagola. 1995. *Numerical methods for the solution of ill-posed problems*, vol. 328. Netherlands: Springer.

11. Paquot, Yvan, François Duport, Anteo Smerieri, Joni Dambre, Benjamin Schrauwen, Marc Haelterman, and Serge Massar. 2012. Optoelectronic reservoir computing. *Scientific Reports* 2: 287.

12. Antonik, Piotr, Marc Haelterman, and Serge Massar. 2017. Brain-inspired photonic signal processor for generating periodic patterns and emulating chaotic systems. *Physical Review Applied* 7: 054014.

13. Soriano, Miguel C., Silvia Ortín, Daniel Brunner, C.R. Laurent Larger, Ingo Fischer Mirasso, and Luıs Pesquera. 2013. Optoelectronic reservoir computing: Tackling noise-induced performance degradation. *Optics Express* 21 (1): 12–20.

14. Soriano, Miguel C., Silvia Ortín, Lars Keuninckx, Lennert Appeltant, Jan Danckaert, Luis Pesquera, and Guy Van der Sande. 2015. Delay-based reservoir computing: Noise effects in a combined analog and digital implementation. *IEEE Transactions on Neural Networks and Learning Systems* 26 (2): 388–393.

15. Antonik, Piotr, Michiel Hermans, François Duport, Marc Haelterman, and Serge Massar. 2016. Towards pattern generation and chaotic series prediction with photonic reservoir computers. In *SPIE's 2016 laser technology and industrial laser conference*, vol. 9732, 97320B.

16. Duport, François, Bendix Schneider, Anteo Smerieri, Marc Haelterman, and Serge Massar. 2012. All-optical reservoir computing. *Optics Express* 20: 22783–22795.

17. Vinckier, Quentin, François Duport, Anteo Smerieri, Kristof Vandoorne, Peter Bienstman, Marc Haelterman, and Serge Massar. 2015. High-performance 142 Chapter V. Towards online-trained analogue readout layer photonic reservoir computer based on a coherently driven passive cavity. *Optica* 2 (5): 438–446.

18. Bauduin, Marc, Quentin Vinckier, Serge Massar, and François Horlin. 2016. High performance bio-inspired analog equalizer for DVB-S2 nonlinear communication channel. In *2016 IEEE 17th international workshop on Signal Processing advances in wireless communications (SPAWC)*, pp. 1–5. IEEE.

# Chapter 6
# Real-Time Automated Tissue Characterisation for Intravascular OCT Scans

One early evening of November 2015, as our group was heading back from the workshop on brain-inspired computing in Besançon, Michiel Hermans casually enquired about any studies or research I have done abroad. My answer was a short "none" at that moment. After a brief discussion of the potential benefits of such an experience for my scientific future, we switched to a different topic that I no longer remember. But the idea of enrolling into an exchange program abroad was planted into my mind, and kept growing until mid-2016, when I started to randomly look for research groups working on machine learning applications to biomedical engineering. The idea of using my knowledge in artificial networks and FPGA programming to improve healthcare was the main motivation at this point.

In July 2016, I picked several research teams in the US and Canada and sent a simple email, with my CV attached. Ten minutes later, I received an interested answer to one of them. This was my first contact with Dr. Thomas E. Milner, a brilliant and very unique professor at the Biomedical Engineering Department (BME) of the University of Texas at Austin (UT). Six months and a few Skype calls later, everything was set up and I was on a plane to Austin, Texas. This chapter tells what came out of this lucky adventure.

## 6.1 Introduction

Atherosclerosis,[1] a disease of the large arteries, is the primary cause of heart disease and stroke [1]. In westernised societies, it is the underlying cause of about 50% of all deaths. It is a progressive disease characterised by the accumulation of lipids and fibrous elements in the large arteries. The early lesions of atherosclerosis consist of subendothelial[2] accumulations of cholesterol-engorged macrophages, called

---

[1] The biomedical background has been adapted from Refs. [1–6].

[2] Subendothelial: beneath the endothelial layer that lines the interior surface of blood and lymphatic vessels.

© Springer International Publishing AG, part of Springer Nature 2018
P. Antonik, *Application of FPGA to Real-Time Machine Learning*,
Springer Theses, https://doi.org/10.1007/978-3-319-91053-6_6

"foam cells". In humans, such "fatty streak" lesions can usually be found in the aorta in the first decade of life, the coronary arteries in the second decade, and the cerebral arteries in the third or fourth decades. Because of differences in blood flow dynamics, there are preferred sites of lesion formation within the arteries. Fatty streaks are not clinically significant, but they are the precursors of more advanced lesions characterised by the accumulation of lipid-rich necrotic debris and smooth muscle cells (SMCs). Such "fibrous lesions" typically have a "fibrous cap" consisting of SMCs and extracellular matrix that encloses a lipid-rich "necrotic core". The whole process of plaque formation is illustrated in Fig. 6.1.

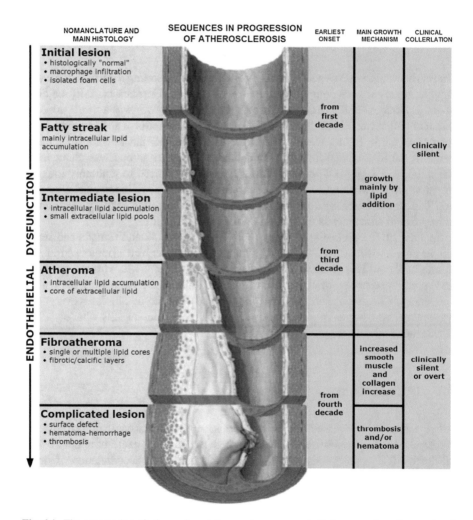

**Fig. 6.1** The progression of atherosclerosis (narrowing exaggerated). Image and caption reprinted from [7]

**Fig. 6.2** Cross section histology (microanatomy) of an artery with a TCFA. The necrotic core, denoted by NC, is covered by a thin fibrous cap, pointed by the arrow. Image reprinted from [8]

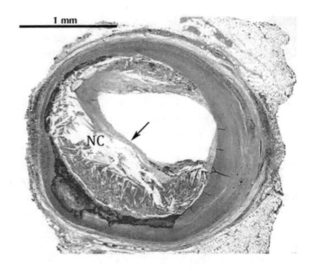

A thin-cap fibroatheroma, or TCFA, is an example of such accumulations, commonly referred to as atheromatous plaques, or simply plaques, in the context of heart or artery matters. It is characterised by a thin fibrous cap, (less than 65 μm in thickness), a large necrotic core, and increased macrophage infiltration [2]. An example of a TCFA is shown in Fig. 6.2. Plaques can become increasingly complex, with calcification, ulceration at the luminal (inner) surface, and haemorrhage from small vessels that grow into the lesion from the media of the blood vessel wall. Although advanced lesions can grow sufficiently large to block blood flow, the most important clinical complication is an acute occlusion due to the formation of a thrombus or blood clot, resulting in myocardial infarction or stroke. Usually, the thrombosis is associated with rupture or erosion of the lesion.

High-resolution visualisation of atherosclerotic plaque morphology is essential for identifying coronary plaques that cause acute coronary events [3, 9]. Several imaging technologies has been developed in the 1990s [4]: intravascular ultrasound (IVUS), magnetic resonance imaging (MRI), and computed tomography (CT). IVUS was widely available for use in interventional cardiology to obtain local structure including reference vessel diameter and localisation of calcium, and to confirm stent deployment. However, the relatively low resolution of IVUS (approx. 100 μm at 40 MHz) prevented visualisation of the fine structures of plaques such as macrophages and thin fibrous caps. MRI had a resolution similar to that of IVUS (approx. 100 μm), and was likewise inadequate to discriminate fine scale plaque features. In the 1990s, coronary artery calcification (CAC) score was examined to estimate the atherosclerotic burden by electron beam or multislice computed tomography (CT). Measurements of CAC score, as quantified by CT, were shown to predict future cardiovascular events. The first use of contrast-enhanced CT to obtain noninvasive coronary angiograms[3] was

---

[3]Angioscopy is a medical technique for visualisation of the interior of blood vessels by inserting a flexible fibre-optic catheter directly into an artery [5].

in 1995, however, the detection of vulnerable plaques prone to rupture on the basis of coronary CT remained challenging. Experimental intravascular methods such as angioscopy had also been proposed to determine plaque vulnerability. Preliminary studies suggested that plaques with a glistening yellow surface, seen by angioscopic examination, were correlated with acute events. However, angioscopy only allowed visualisation of the surface of plaques and additionally required extended periods of blood flow occlusion, which carried risk of ischaemia.

Intravascular optical coherence tomography (OCT) is an established medical imaging technique that produces high-resolution cross-sectional imaging of biological tissues by measuring the magnitude and "echo time delay" of backscattered light, instead of ultrasound in IVUS [6]. Since the speed of light is much faster than that of sound, an interferometer is required to measure the backscattered signal. The interferometer uses a fibre-optic coupler similar to a beam splitter, which directs half of the beam to the measurement arm (tissue) and the other half to the reference arm. The light source used by intravascular OCT has a near-infrared spectrum, with wavelengths ranging from 1250 to 1350 nm. Figure 6.3 shows a simplified scheme of an intravascular OCT setup. The axial resolution of current OCT systems ranges from 10 to 15 μm, which allows detailed visualisation of biological structures at a micro-scale level. Cross-sectional images, called B-scans (top panels in Fig. 6.4), are created from a series of axial scans, called A-scans (bottom panel in Fig. 6.4).

Intravascular OCT allows to distinguish the three types of plaque (fibrous, fibrocalcific and lipid-rich) with sensitivity and specificity up to roughly 90% [3]. However, the characterisation process requires human interpretation of the OCT images, guided by a series of established criteria. In 2016, the team lead by Dr. Milner achieved the

**Fig. 6.3** Intravascular OCT creates an image by sending a light beam into the tissue and measuring the intensity of the reflection by comparing it with a reference beam. Image reprinted from [10]

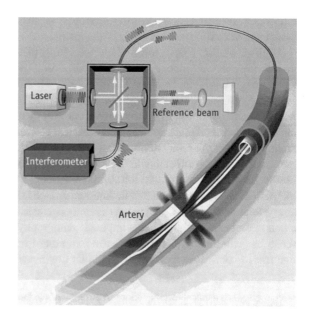

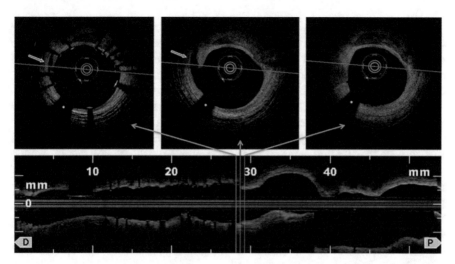

**Fig. 6.4** Examples of cross-section images (B-scans, top panels), obtained from intravascular OCT axial scans (A-scans, bottom panel shows one example) at three different locations in the artery, indicated by blue lines. The white asterisk in the top panels indicates the OCT catheter. Images reprinted from [11]

first results on automated plaque characterisation from OCT images using artificial intelligence. The preliminary results were published in [12, 13], and a journal paper was on the way at the moment of writing these lines. The authors reported sensitivity and specificity comparable to the previous work [3], with the advantage of proposing a fully-automated process, thus excluding the uncertainty due to the human factor.

The above study was carried out as follows [14]. IV-OCT imaging was conducted on 25 human hearts collected within 24 h of death. The age at death was $65 \pm 11$ years. Imaging was conducted on 54 coronary arteries. Coronary arteries were dissected from the heart and placed onto a custom metal device with millimetre markers to assist with co-registration with histology. Arteries were pressurised at 100 mm Hg with saline solution prior to insertion of a St Jude dragonfly OCT imaging coronary catheter. Automated pull was performed at 50 mm/s and the images stored for later analysis.

For the histology, the arteries were individually radiographed and decalcified overnight if excessive calcium was present. The arterial segments were sliced into 2–3 mm thick rings and further processed on a Tissue-Tek Vacuum Infiltration Processor (Sakura Finetek USA, Torrance, CA) for standard paraffin-embedded sections. An average of 25 rings were generated from each artery. Serial tissue sections (5 μm thick) were cut at 150-μm intervals and stained with hematoxylin and eosin (H&E), modified Movat's pentachrome, Von Kossa. Anti-CD68 (Dako North America, Inc, Carpinteria, CA) and anti-$\alpha$-smooth muscle cell-actin (Sigma-Aldrich, St. Louis, MO) antibodies were used in immunohistochemical studies to identify macrophages and smooth muscle cells, respectively. Histology rings were then matched to respective IV-OCT frames and used as ground truth for the automated method.

The major drawback of the automated method is the runtime. The analysis of one B-scan image takes typically 6–8 h, while processing the entire pullback (containing up to 300 images) requires 3 to 4 days. Such delays are detrimental for practical clinical applications. As most image processing techniques, the automated plaque characterisation is a very computationally intense task. However, it could be significantly sped up by using dedicated electronics, such as a FPGA chip, to do the hard work, instead of a computer. And this is where I come up on the stage.

The method developed by Dr. Milner's team [12, 13] can be split in two stages. The first consists in gathering the most relevant information from the image, something more elaborate and useful than just the raw value of each pixel. In the field of image processing, this is commonly called "feature extraction", where a feature is a particular characteristic of the image, computed following a certain algorithm. A few examples of well-known features of an image are the contrast and the homogeneity. In the second the stage, the image features are processed by a classifier, trained to distinguish the three types of plaque. In this work, a simple feedforward artificial neural network (see Sect. 1.1.2) was used.

Originally, my project was to implement the neural network on the FPGA to increase its processing speed. After the first brainstorming at UT, the speed-up of the feature extraction process became the primary objective, as this stage was the most time consuming. By the end of my internship, we successfully implemented both stages on the FPGA with significant speed gains, as will be described below. Section 6.2 covers the feature extraction and Sect. 6.3 details the neural network design. Both sections have similar structures, with a theoretical introduction, an overview of the FPGA design, followed by the results and perspectives.

For this project, the UT team purchased a brand new FPGA board—the Xilinx VC707 with a Virtex-7 chip, depicted in Fig. 6.5. The new 7 family was released in

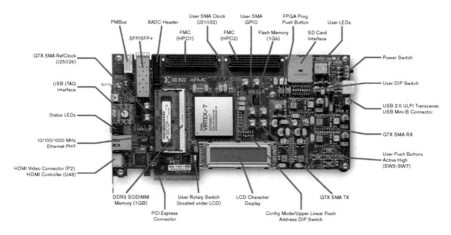

**Fig. 6.5** Xilinx VC707 evaluation board featuring by the Virtex-7 XC7VX485T-2FFG1761C FPGA. The attributes of the board, such as connectivity ports, LCD display, LEDs and buttons, are very similar to the ML605 board. Reprinted from Xilinx website

2012 as an upgrade of the existing Virtex-6 family. The internal components (logic, registers, DSP slices, block memory) become faster and come in greater numbers as a results of an improved manufacturing technology (28 nm versus 40 nm for the 6 family). For instance, our Virtex-7 XC7VX485T contains 2800 DSP slices, in comparison to 768 in the Virtex-6 I was previously using. But most importantly, the new 7 family comes with a whole new software—the Vivado Suite, that completely replaces the ISE Design Suite, with no forward or backward compatibility. The adaptation to the new software requires some time, but the effort is fully justified, as Vivado is much better than ISE on multiple levels.

## 6.2  Feature Extraction

Let us start with the first project accomplished at the University of Texas.

### 6.2.1  GLCM Features

The classification or characterisation of pictorial data requires a set of features accurately describing the information [15]. In a search for meaningful features, it is only natural to look toward the types of features which human beings use in interpreting pictorial information. Spectral, textural, and contextual features are three fundamental pattern elements used in human interpretation of colour images. Spectral features describe the average tonal variation on various bands of the visible portion of an electromagnetic spectrum, whereas textural features contain information about the spatial distribution of tonal variation within a band. Contextual features contain information derived from blocks of pictorial data surrounding the area being analysed.

Gray-level co-occurrence matrix (GLCM) texture method, also known as co-occurrence matrix or distribution, is a way of extracting second order statistical texture features from grey-level images [16]. A GLCM is a $N_g \times N_g$ matrix, where $N_g$ is the number of quantised grey levels in the image. The matrix elements $p(i, j)$ contain the number of pairs of pixels with grey levels $i$ and $j$, respectively, that satisfy a spatial relationship defined by the user. A simple example of a GLCM matrix is presented in Fig. 6.6. The spatial relationship can be expressed in terms of either horizontal and vertical offsets, or angle-distance pairs $(\theta, d)$. For example, the "one pixel to the right" relationship, considered in Fig. 6.6, corresponds to +1 horizontal offset and 0 vertical offset, which is equivalent to an angle of $\theta = 0°$ and a distance of $d = 1$. In-depth discussion of GLCM matrices and their properties lies beyond the scope of the present thesis, but interested readers can find more information in [15, 17].

In practice, the GLCM matrix is computed iteratively, as illustrated in Fig. 6.6. In this example a $4 \times 5$ image holds pixels with 8 greyscale levels. Thus, the starting point of the algorithm is a $8 \times 8$ zero matrix. The size of the image does not affect the

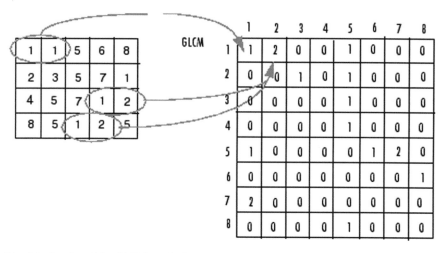

**Fig. 6.6** Creation of the GLCM matrix from a simple 4 × 5 image (left-hand side). The greyscale pixels hold 8 grey levels, which results in a 8 × 8 GLCM matrix (right-hand side). The matrix contains the information about pixel pairs satisfying a certain spatial relationship. In this example, "one pixel to the right" relation is considered. A few examples are highlighted with gray ellipses. The row number corresponds to the value of the reference pixel (the left one in the ellipse), and the column number gives the greyscale level of the neighbour pixel (the right one). Matrix elements hold the number of particular greyscale level relations found within the image. The matrix is filled iteratively, by checking each individual pixel of the image, except for the border ones, that do not have a neighbour satisfying the relation (in this example, the right-most column of the image). Reprinted from Matlab GLCM documentation

size of the matrix. Then, the entire image is scanned, pixel by pixel. The algorithm considers the relation between two pixels at a time, called the reference and the neighbour pixel. Each pixel of the image becomes the reference pixel in turn, starting from the upper left corner and proceeding to the lower right. The neighbour pixel is determined by the spatial relationship, defined above. In this example, the neighbour pixel is the immediate neighbour to the right. Note that pixels from the right column of the image are not accounted for, as they do not have a neighbour on the right. For each pixel pair considered, the GLCM matrix cell $(x, y)$ is incremented by 1, where $x$ and $y$ are the greyscale levels of the reference and neighbour pixels, respectively.

A large number of textural features can be extracted from a GLCM matrix [15]. The following five features were used in [12, 13]. The goal of my project was to rapidly compute these five features on the FPGA. The equations and explanations are adapted from [16].

- The energy (also known as angular second moment)

$$\text{Energy} = \sum_{i=0}^{N_g-1} \sum_{j=0}^{N_g-1} p(i, j)^2 \qquad (6.1)$$

is a measure of homogeneity of an image. A homogeneous texture will contain only a few grey levels, so that GLCM will have a few but relatively high values of $p(i, j)$. Thus, the sum of squares will be high. Therefore, when the image is homogeneous, the energy will have high values.

• The homogeneity (also known as inverse difference moment)

$$\text{Homogeneity} = \sum_{i=0}^{N_g-1} \sum_{j=0}^{N_g-1} \frac{1}{1 + (i - j)^2} p(i, j) \tag{6.2}$$

measures the local homogeneity of an image. The incidence of co-occurrence of pixel pairs is enhanced when they are close in grey-scale value and thus increases the homogeneity value. Because of the weighting factor $(1 + (i - j)^2)^{-1}$, it will get small contributions from inhomogeneous areas $i \neq j$. The result is a low homogeneity value for inhomogeneous images, and a relatively higher value for homogeneous images.

• The contrast

$$\text{Contrast} = \sum_{i=0}^{N_g-1} \sum_{j=0}^{N_g-1} (i - j)^2 p(i, j) \tag{6.3}$$

is a measure of the local variations presented in an image. This measure of contrast favours contributions from $p(i, j)$ away from the diagonal. If there is a large amount of variation in an image, the contrast will be high.

• The entropy

$$\text{Entropy} = - \sum_{i=0}^{N_g-1} \sum_{j=0}^{N_g-1} p(i, j) \log [p(i, j)] \tag{6.4}$$

measures the randomness of the image texture (intensity distribution). Entropy is the highest when all the probabilities $p(i, j)$ are equal, and smaller when the entries in $p(i, j)$ are unequal. Therefore, a homogeneous image will result in a lower entropy value, while an inhomogeneous (heterogeneous) region will result in a higher entropy value.

• The maximum probability is simply the highest value found in the GLCM matrix.

Texture features can be computed globally—that is, one feature value for the entire image, or locally—one value for each pixel or region of the image. Since the task here is to distinguish different tissues within an image, the features are computed locally [12, 13]. A small square region, called a window, is defined around each pixel of the image. The GLCM features are computed for each such window and stored in tables (each feature gives a separate table). Since a window cannot be defined for each pixel of the image—pixels close to the borders do not have enough neighbours on the sides—the resulting table is smaller than the image. For instance, the homogeneity of a $100 \times 100$ image, computed with a $11 \times 11$ window, is a $90 \times 90$ table. Each value of the table quantifies the homogeneity of a window around the corresponding pixel of the image, e.g. the (1, 1) cell of the table (upper left-most) corresponds to the homogeneity of the window around the pixel (6, 6).

The choice of the spatial relationship, or angle-distance pair $(\theta, d)$, plays an important role. Since it defines how the GLCM matrix is computed, the resulting features provide different textural characteristics of the image. Intuitively, short-distance relationships provide features relevant on small-scale, while long-distance relationships give a more wide-scale description of the image. Full characterisation of image textural properties may require combining different relationships. This means that for each window, multiple GLCM matrices are computed (with different angle-distance pairs), together with the corresponding features. For instance, one can compute three homogeneity tables for angle-distance pairs $(0°, 1)$, $(0°, 2)$ and $(90°, 3)$.

## 6.2.2  Methods

The input image, typically $666 \times 662$ pixel, is stored in FPGA BRAM (see Sect. 1.3.3). The 6-bit pixels set the size of the GLCM matrix at $64 \times 64$. The matrix is computed for a $23 \times 23$ pixel window. In [12, 13], 24 angle-distance pairs are considered, meaning that 24 GLCM matrices are computed for each window of the image. In this proof-of-principle implementation, we only evaluate two angle-distance pairs. To speed up the process, we compute one GLCM matrix per window, and exploit the central symmetry to account for two angle-distance pairs $(\theta, d)$ and $(\theta + 180°, d)$.

The full algorithm, described in the previous section, can be split in two procedures: (1) the computation of the GLCM matrix for each window, centred on a given pixel and (2) the computation of the five features associated with each matrix. The following paragraphs describe how these two steps are carried out by the FPGA.

The process starts in the upper-left corner of the image. The window is gradually shifted by one pixel to the right, until it reaches the border of the image. Then, the moving window returns back to the left border and moves down by one pixel. That is, the image is processed line-wise, until the lower-right corner is reached. To save BRAMs, the window under consideration is not duplicated. Instead, two counters are defined to restrict the pixels within the window of interest.

The GLCM matrix is computed for one angle-distance pair. By design, the matrix is symmetric: each pair is accounted for twice, thus exploiting the central symmetry. One GLCM matrix is built for each window. This process is carried out in two different ways, depending on the window position. The first (left-most) window on each line is fully processed. That is, each pixel pair satisfying the relation is considered. The following windows on the same line are processed in a more efficient, incremental way, following the idea proposed by Austin McElroy from UT. The algorithm subtracts the pixel pairs from the left-most column (those that were left behind after the window shift) from the GLCM matrix, and adds the new ones from the right-most column. This allows for a significant speed up of the process.

Since analysing one pixel pair requires 5 clock cycles, the full window requires $5 \cdot 22^2 = 2420$ clock cycles to be processed. On the other hand, the shifting method allows to update the GLCM matrix in 220 clock cycles.

Once the GLCM is ready, a trigger is sent to the computational modules that evaluate the five features in parallel (simultaneously). Four of the five features require the same series of computations—an element-wise matrix multiplication, followed by a summation of all the elements. For this reason, a generic module has been implemented that can be used for the four features. The element-wise product of two $64 \times 64$ matrices, and the subsequent summation, are carried out in a semi-parallel way by 64 DSP slices. The process is completed in 70 clock cycles. This strategy is a trade-off between the execution speed and optimal FPGA resource utilisation.[4] For homogeneity and contrast, the multiplication requires a matrix of constant coefficients (see Eqs. 6.2 and 6.3). These are defined by $i$ and $j$ coordinates in the GLCM matrix and can be computed in advance. They are pre-computed on the computer and hard-coded into the FPGA design as constants to save resources. The entropy computation involves the log function (see Eq. 6.4). It is implemented with a 12-bit look-up table, stored in the FPGA memory. The 12-bit resolution was chosen as a trade-off between the precision of the results and the FPGA resource utilisation. The only feature computed differently is the maximum probability. Its evaluation does not require complex computation and boils down to simple number comparison.

### 6.2.3   Operation Principle

The feature extraction procedure is schematised in Fig. 6.7. The FPGA board is connected to two computers. The main computer, running Matlab, controls the operation of the design and sends the image to the chip. A second computer, running Xilinx Vivado software, is used to configure the chip, monitor and debug its behaviour. Since the features are extracted through the `debug` module (as will be explained below), the computed features appear on the second computer.

The design can be set in one of the four pre-defined states:

**idle**:   The design does nothing and all internal variables are reset to default values.
**setup**:   The BRAMs are ready for receiving data.
**comp**:   The image is processed, following the methods described above.
**bramcheck**:   Debug mode, designed to check the FPGA memory. All BRAMs are programmed to burst out their contents.

Switching the FPGA state is done by executing specific commands in Matlab. At first, the FPGA is set to `setup` mode and the image is written into the memory. The `bramcheck` mode can then be activated to make sure the data has been

---

[4]Similar to BRAMs, DSP slices are the limiting resource on a FPGA chip.

Matlab                          VC707                          Vivado

**Fig. 6.7** Scheme of the feature extraction procedure on the FPGA board. The chip receives the image from a computer running Matlab, but the resulting features are collected on a laptop through the Xilinx Vivado software

written correctly. The computations are launched by switching the `comp` mode. The computed features are recorded on the debug computer.

### 6.2.4  FPGA Design

The FPGA design is written in standard IEEE 1076-1993 VHDL language [18, 19] and implemented with Xilinx Vivado 2016.4 (64-bit). The simplified schematics of the design is shown in Fig. 6.8. Rectangular boxes depict modules (entities), and the lines represent connections between them.

The operation of the FPGA is controlled from the computer via a simple custom protocol through the Universal Asynchronous Receiver/Transmitter (UART) connection. Data and commands are encoded into 8-bit words. The 2 most significant bits define the word type (command or data), while the 6 least significant bits contain

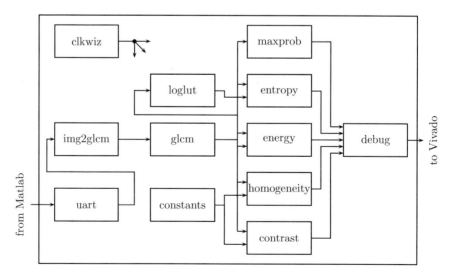

**Fig. 6.8** Simplified schematics of the FPGA design. Modules (entities) are represented with rectangular boxes. Their functions are described in the text

the data or the command itself. The only data the FPGA is supposed to receive is the image to process. Since each pixel has 64 levels of grey, it can be encoded as a 6-bit values. The connection speed is set to 115200 bit/s.

The following gives a short description of each module in the design.

**clkwiz**: Master clock managing module, it converts the onboard 200 MHz clock into a lower-frequency clock, used to drive the design. This design is driven by a 140 MHz clock. This is the highest frequency that allows to meet timing closure.

**uart**: Main communication module that interfaces the FPGA with Matlab, running on the control computer. It takes care of detecting the incoming bytes and interpreting the commands and data received.

**img2glcm**: The image to process is stored in BRAMs within the `img2glcm` module: 662 blocks of 18 kbit are used, each BRAM holds one line of the image. The module handles the moving window and computes the GLCM.

**glcm**: The GLCM is contained in 64 BRAMs, one per line. The matrix is filled by the `img2glcm` module, while the reading of the GLCM during the `comp` stage is handled by this module. That is, it outputs one column (64 values) per clock cycle during 64 clock cycles. The values are fed into 4 identical modules, computing the energy, homogeneity, contrast and entropy, as well as a different module computing the maximum probability (see below).

**loglut**: A 12-bit look-up table, implemented as ROM into the FPGA distributed memory. This allows to take the logarithm of the full 64-value column in one clock cycle, that is, compute 64 logarithms simultaneously.

**constants**: This module consists of two $64 \times 64$ matrices, also implemented as distributed ROM, containing the constant coefficients used for computation of homogeneity and contrast.

**entropy, energy, homogeneity, contrast**: Four identical generic modules, computing an accumulated element-wise product of two matrices. Each module employs 64 DSP slices, performing 64 multiplications (1 full column) per clock cycle. The module has two 64-value-wide inputs. The incoming values are multiplied and accumulated line-wise for 64 clock cycles, then the resulting 64 values (the sums over each line) are added in pairs in 6 clock cycles, giving the final result.

**maxprob**: This modules computes the only feature not involving matrix multiplication. Data is received in the same manner: one GLCM column per clock cycle. The process first selects the maximum value in each row of the matrix (in 64 clock cycles), then compares the rows in pairs to get the overall maximum (in 6 clock cycles).

**debug**: This module contains Integrated Logic Analyser (ILA) cores, meant to debug and monitor the internal signals. In the present version of the design, the final features values are extracted through this module.

The arithmetic operations computed by the FPGA are performed on real numbers. However, the chip is a logic device, designed to operate bits. The performance of the design thus highly depends on the bit-representation of real numbers, i.e. the precision. The main constraint comes from the DSP slices, capable of multiplying a 25-bit signed integer by a 18-bit signed integer. For this reason we designed the mathematical operations for optimal utilisation of the resolution available. Since all values are positive, they can be treated as unsigned. All values are tuned to lie within a [0, 1] interval, so that they could be represented as 16-bit integers, with 16 bits for the decimal part. To this end, we artificially divided the coefficients for contrast by an empirical value of 4096, so that the final value within the FPGA is strictly inferior to one. However, the actual contrast should be multiplied by 4096 afterwards. We applied the same trick to the entropy by dividing the logarithm look-up table values by a factor of 16.

### 6.2.5   Results

The present design, driven by a 140 MHz clock, processes one $662 \times 666$ pixel image for 2 angle-distance pairs (see Sect. 6.2.1) in 0.893 s. That is, it should complete the same image for the full set of 24 pairs (see Sect. 6.2.2 and [12, 13]) in 10.72 s.

The main bottleneck of this implementation is the slow UART connection. Transferring the full image to the FPGA takes approximately 38 s. Moreover, the connection is too slow to send the computed features back to the control computer. This is why they have to be extracted through the `debug` module on the debug computer.

A sample containing 16384 values for each feature was extracted to evaluate the FPGA precision. The mean and maximum relative errors are given in Table 6.1. The error is less than 1% for all features except for the Entropy, where the higher error is induced by the low-resolution log function.

The FPGA resource utilisation is given in Table 6.2. The high BRAM utilisation is due to large ILA cores used for debugging.

**Table 6.1**   FPGA precision (relative errors)

|  | Mean ($\times 10^{-3}$) | Maximum ($\times 10^{-3}$) |
|---|---|---|
| Energy | 8.25 | 9.85 |
| Homogeneity | 4.29 | 4.37 |
| Contrast | 3.88 | 5.02 |
| Entropy | 10.20 | 11.69 |
| Max Probability | 0 | 0 |

**Table 6.2**   FPGA resource utilisation summary

|  | LUT | FF | BRAM 18 kb | DSP |
|---|---|---|---|---|
| Used | 46650 | 27533 | 1029 | 256 |
| Available | 303600 | 607200 | 2060 | 2800 |
| Utilisation (%) | 15.37 | 4.53 | 49.95 | 9.14 |

### 6.2.6 Perspectives

The main bottleneck (slow UART connection) can be solved with a much faster Gbit Ethernet or PCI Express connection. The image transfer should then take no longer than a fraction of a second, and the FPGA could also send the features back to the computer in real time. The design performance can be further improved by optimising both the methods and the hardware.

The methods could be simplified as follows. Reducing the number of angle-distance pairs is the first idea that comes to mind: are all 24 pairs really necessary? The FPGA runtime is proportional to this number. In order to define the most relevant features, one should apply several well-established methods in machine learning, commonly called feature selection or dimensionality reduction. Their overview lies beyond the scope of this thesis. The second idea suggests reducing the number of grey levels in the image, currently at 64. This would reduce the size of the GLCM matrix. A decrease down to 32 levels of grey would help optimising the hardware through multi-threading (more on that in the next paragraph). Ideally, the number should be 16 or lower, so that the GLCM could be stored in the FPGA distributed memory. This would allow to speed-up the read and write functions, currently taking 5 clock cycles per GLCM cell.

On the hardware side, the relatively slow operation of BRAMs requires 5 clock cycles to update one element of the GLCM matrix. Moving the GLCM from the BRAMs to the distributed memory (this would be possible with 16 grey levels or less) would reduce the latency down to 2–3 clock cycles. Since the FPGA resource usage is relatively low, multi-threading becomes an interesting possibility. Current design has one "block" building the GLCM and one another computing the features. These blocks do not use all the resources on the device, so that one could add more such block. This is similar to creating multiple CPU cores processing the image in parallel threads.

In principle, each one of these points could speed up the design by a factor of 2 or 3. Combining them together should allow one to reduce the runtime by an order of magnitude and reach real-time feature extraction.

## 6.3 Artificial Neural Network

The neural network was the second project accomplished after the feature extraction implementation.

### 6.3.1 Network Structure

The neural network is used to characterise the tissue types within intravascular OCT B-scans (see Sect. 6.1). The inputs to the network are the textural features, extracted from the image. In the previous Sect. 6.2 we have covered five GLCM features that we implemented on FPGA for high-speed computation. However, many more additional features—up to 300—are used for classification in [12, 13]. Their discussion lies beyond the scope of the present thesis, and we refer the reader to papers [12, 13] for a complete list of features. During my internship, we only focused on the GLCM features that were the most challenging to compute on the FPGA. The remaining features will be implemented in the future work (more on that in Sect. 6.4). The goal of the neural network is to distinguish between the three types of plaque (fibrous, lipid and calcium). To this end, the network is trained to characterise each pixel based on its features, that is, to output {0, 0.5, 1} for each type of tissue. The choice of these values will be explained below.

To emulate the same network on FPGA, we had to reverse-engineer several Matlab functions, since some network characteristics are very poorly documented. Luckily, a "black box"-like neural network in Matlab (see Fig. 6.9) can be converted into a precise and explicit Simulink model (see Fig. 6.11), that reveals its internal structure.

The basic structure of the Matlab neural network for pattern classification (created with the `patternnet` function) is depicted in Fig. 6.9. The network is configured with 300 inputs (equal to the number of image features [12, 13]), one hidden layer with 200 neurons and one output neuron. From the perspective of machine learning, this is a simple and relatively small feedforward neural network, as discussed in Sect. 1.1.2.

The inputs are contained in a $300 \times L$ matrix, where each column is one 300-dimensional input, and $L$ is the number of inputs. Each input is multiplied by the input weights, contained in a $200 \times 300$ matrix. This matrix can be seen as a set of 300 weights (a row) for each one of the 200 neurons in the hidden layer. The multiplication thus assigns a weighted sum of the 300 inputs to each neuron. Note that each neurons gets a different mapping, and that all inputs are mapped to all neurons (there are no null inputs weights). In addition to the weighted sum of inputs, each neuron receives a constant bias from a $200 \times 1$ vector. The biases are different for each neuron, but the same bias is added with each new input.

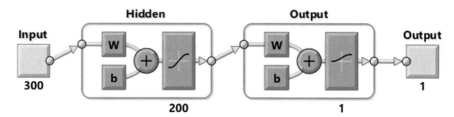

**Fig. 6.9** Simplified structure of the Matlab neural network. The 300-dimensional input is mapped into 200 neurons of the single hidden layer. The output layer computes the value of a single output neuron as a weighted sum of the hidden neurons. Although the input and output weights and biases are marked with the same symbols—W and b—their values differ (see main text for details)

**Fig. 6.10** Matlab `tansig`
and `logsig` activation
functions. Note the different
ranges and slopes of the two
functions

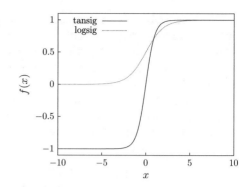

The sum of weighted inputs and the bias for each neuron is then processed by the activation function, set to the default `tansig` in this case. The `tansig` function, plotted in Fig. 6.10 and given by

$$\text{tansig}(x) = \frac{2}{1 + e^{-2x}} - 1, \tag{6.5}$$

is a Matlab implementation of the hyperbolic tangent sigmoid transfer function. Formally, it is equivalent to the standard hyperbolic tangent, with the advantage of faster computation, as it only contains one exponential function. Note, however, that different ways of computing the hyperbolic tangent can give rise to very small numerical differences. This function is a good trade-off for neural networks, where speed is important and the exact shape of the transfer function is not [20]. The output of the function gives the states of the 200 neurons of the hidden layer.

In the output layer, the states of the hidden neurons are used to evaluate the single output neuron. This is done by computing another weighted sum, now combining the 200 neurons with 200 readout weights. A bias is also added, before feeding the result through the `logsig` function—another Matlab activation function [21], plotted in Fig. 6.10 and given by

$$\text{logsig}(x) = \frac{1}{1 + e^{-x}}. \tag{6.6}$$

The output of the `logsig` function is the output of the neural network for the current input.

The operation of the neural network can be summarised by the following equation

$$y = \text{logsig}\left(W \cdot \text{tansig}\left(I \cdot x + b_i\right) + b_r\right), \tag{6.7}$$

where $x$ is the 300-dimensional input, $y$ is the one-dimensional output, $W$ and $I$ are the readout and input weights, respectively, and $b_i$ and $b_r$ are the input and readout biases, respectively.

However, Matlab does more data processing "under the hood", so one has to dig into the Simulink model to figure out what happens to the inputs. The expanded model of the network is depicted in Fig. 6.11. Figures 6.12 and 6.13 illustrate the internals of the "Layer 1" (hidden layer) and "Layer 2" (output layer) blocks. Their operation has been described above. However, "Process Input 1" and "Process Output 1" blocks have been left in the shadow so far. They perform basic data pre- and post-processing, that can dramatically change the inputs that the neural network actually receives, and the outputs that the user gets. The internal structure of these blocks is shown in Figs. 6.14 and 6.15. The pre-processing consists of the following functions:

removeconstantrows:   This function removes constant rows, that is, inputs
    to the network that are constant across all the timesteps; such inputs are either
    useless to the network, or can be incorporated into the input and output biases.
mapminmax:   This function scales the input rows (individually) so that the values
    lie in the $[-1, +1]$ interval.

The output post-processing consists of the same functions, performing the reverse operations. To simplify the FPGA implementation, we pre-process the data (image features) in Matlab, before feeding it into the FPGA, and configure the network so that the targets lie within the $[0, +1]$ interval (the output range of the logsig function), so that no post-processing of the outputs is necessary. Thus, the network is trained to output $\{0, 0.5, 1\}$ for the three types of plaque {fibrous, lipid, calcium}.

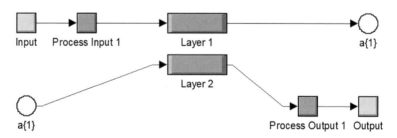

**Fig. 6.11** Complete Simulink model of the Matlab neural network. Layers 1 and 2 are the hidden and output layers, respectively. The inputs and the output are pre- and post-processed by additional blocks "Process Input 1" and "Process Output 1", respectively

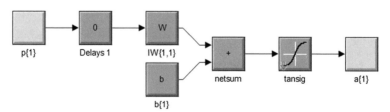

**Fig. 6.12** Simulink model of the hidden layer. The inputs p{1}, weighted by the input weights IW{1,1}, are added to the input bias b{1} and processed by the tansig transfer function. The delay block is not used here

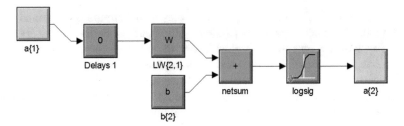

**Fig. 6.13** Simulink model of the output layer. The hidden neurons a{1}, weighted by the readout weights LW{2,1}, are added to the output bias b{2} and processed by the logsig transfer function. The delay block is not used here

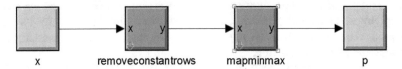

**Fig. 6.14** Simulink model of the input pre-processing. The user-defined inputs x are processed by two functions that remove constant inputs (rows) and scale the remaining rows into the [−1, 1] interval

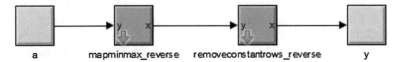

**Fig. 6.15** Simulink model of the output post-processing. It performs the reverse operations of the input pre-processing

### 6.3.2 Methods

The training of the network is performed on the computer, and the optimal weights are uploaded into the FPGA. The values are stored in BRAMs, except for the single-value output bias ($b_r$ in Eq. 6.7) that is stored in registers.

Three BRAMs are allocated for the inputs, input weights and output weights, respectively. Note that the first block (for inputs) will not be required in a unified design, comprising the feature extraction process. That is, once this neural network implementation is combined with the previous design computing the GLCM features, there will be no need to store the inputs, as they would be processed on the fly (more on that in Sect. 6.4).

To simplify the operation of the design and save FPGA resources, the input biases ($b_i$ in Eq. 6.7) are treated as additional input weights, multiplied by a fictional 298-th input equal to 1. This requires two additional BRAMs, but saves a significant amount of logic and registers.

The inputs and input weights are continuously read from the BRAMs and multiplied by the DSP slices. With 200 parallel multiplications, performed by 200 slices at a cadence of one operation per clock cycle, the design requires 298 clock cycles to map one 297-dimensional input and the input biases into the hidden layer. Once this is done, the resulting values are saved into registers, and a trigger is generated to activate the `tansig` function. Meanwhile, the DSP slices keep on processing data, now multiplying the next input by the same input weights. This design technique allows to exploit the intrinsic parallelism of FPGA chips and get more computations done in fewer clock cycles.

The `tansig` function is implemented with a 12-bit look-up table. It registers the input with the input trigger and produces the output on the next clock cycle, together with an output trigger. The latter launches the last single DSP slice, taking care of multiplying and accumulating the 200 hidden neurons with the 200 output weights. The process requires 200 clock cycles. Once the output is computed, another trigger is sent to the `logsig` function.

The `logsig` function is implemented in the same way as the `tansig` function. The output bias is added directly to the input of the function. The output value is produced on the next clock cycle, which completes the processing of one neural network input.

In summary, the design consists of two parallel processes: the first computes the $I \cdot x + b_i$ part of Eq. 6.7 and requires 298 clock cycles, while the second computes the remaining $\text{logsig}(W \cdot \text{tansig}(\ldots) + b_r)$ in 204 clock cycles. This results in a 298 clock cycles throughput, with a 204 clock cycles latency. In simple words, this means that the design produces an output every 298 clock cycles, but there is an additional 204 clock cycles latency between a given input and its corresponding output. Such latency is negligible on the long run.

### 6.3.3  Operation Principle

The neural network implementation on FPGA is schematised in Fig. 6.16 and is very similar to the feature extraction scheme presented in Sect. 6.2. The main (and only) difference is that the design receives the features and the weights (instead of an image) and outputs the plaque types, or classes (instead of the features).

### 6.3.4  FPGA Design

The FPGA design is written in standard IEEE 1076-1993 VHDL language [18, 19] and implemented with Xilinx Vivado 2016.4 (64-bit). The simplified schematics of the design is shown in Fig. 6.17. Rectangular boxes depict modules (entities), and the lines represent connections between them.

**Fig. 6.16** Scheme of the neural network processing on the FPGA board. The board receives the features from a computer running Matlab, but the resulting plaque types (classes) are collected on a laptop through the Xilinx Vivado software

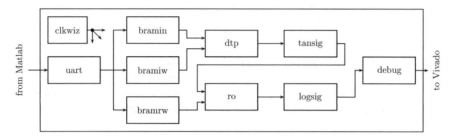

**Fig. 6.17** Simplified schematics of the FPGA design. Modules (entities) are represented with rectangular boxes. Their functions are described in the text

The operation of the FPGA is controlled from the computer via a simple custom protocol through the UART connection. Data and commands are encoded into 24-bit words. The 2 most significant bits define the word type (command or data), while the 16 least significant bits contain the data or the command itself. The connection speed is set to 115200 bit/s.

The following gives a short description of each module within the design.

**clkwiz**: Master clock managing module, it converts the onboard 200 MHz clock into a lower-frequency clock, used to drive the design. This design is driven by a 180 MHz clock. This is the highest frequency that allows to meet timing closure.

**uart**: Main communication module that interfaces the FPGA with Matlab, running on the control computer. It takes care of detecting the incoming bytes and interpreting the commands and data received.

**bramin, bramiw, bramrw**: These similar modules contain blocks of RAM required to store the inputs and the network weights. Specifically, 298 BRAMs are used to store the inputs, 200 BRAMs for the input weights and 1 for the output weights. The writing into BRAMs is performed through a shared bus. The design contains a `target` signal, controlled from Matlab, indicating the destination BRAM type.

**dtp**: The generic dot product module computes the product of two vectors of arbitrary length. This module is replicated 200 times in the design, each one computing the weighted input to one neuron. These modules function continuously, in parallel with the `tansig`, `ro` and `logsig` modules (as explained in Sect. 6.3.2).

**tansig, logsig**: Two 12-bit look-up tables, implemented as ROMs into the FPGA distributed memory. They allow to process as many inputs as necessary in parallel.

The input is registered on the input trigger and the output is ready on the next clock cycle.

**ro**:    The readout module consists of another dot product module, combined with a serialiser process, converting the 200 concurrent neuron values into a stream with one value per clock cycle. It is launched by the output trigger from the `tansig` module. Upon completion, it generates its own trigger that activates the `logsig` function.

**debug**:    This module contains Integrated Logic Analyser (ILA) cores, meant to debug and monitor the internal signals. In the present version of the design, the final neural network outputs are extracted through this module.

Similar to the feature extraction design, the precision is of particular importance. However, the main difference here is that all number can be either positive or negative. Therefore, the inputs and weights are treated as 16-bit signed. Since some input features may exceed 1, we allocate 12 bits to the decimal part, 3 for the integer part and 1 for the sign. This allows to prevent overflow in all computations.

### 6.3.5   Results

The present design, driven by a 180 MHz clock, processes one network input in 66 μm. Therefore, it should complete the full $666 \times 666$ pixel image (with 23-pixel window) in 0.69 s.

Again, the main bottleneck is the slow UART connection. Uploading the inputs and the weights onto the FPGA takes approximately 60 s. Here as well, the connection is too slow to send the outputs back to the computer. However, uploading the inputs will no longer be necessary in the final version of the design (more on that in Sect. 6.4), and the slow connection shall be replaced by a high-speed PCI Express bus (see Sect. 6.2.6).

The design was tested on a 600-input sample of the original data. The neural network outputs were extracted through the `debug` module to evaluate the FPGA precision. The mean relative error is inferior to 1%.

The FPGA resource utilisation is given in Table 6.3. The high BRAM utilisation is partially due to large ILA cores used for debugging.

**Table 6.3** FPGA resource utilisation summary

|                 | LUT    | FF     | BRAM 18 kb | DSP  |
|-----------------|--------|--------|------------|------|
| Used            | 21065  | 29609  | 1028       | 201  |
| Available       | 303600 | 607200 | 2060       | 2800 |
| Utilisation (%) | 6.94   | 4.88   | 49.90      | 7.18 |

## 6.4 Conclusion

In this work, we created two proof-of-principle FPGA prototypes to demonstrate the possibility of high-speed automated plaque classification. The two processing stages—feature extraction and classification by a neural network—could be completed in less than 11 s and 0.7 s respectively. This represents an up to $10^3$ gain in processing speed (compared to a high-end PC), that has been achieved with the intrinsic parallelism and dedicated usage of the FPGA resources.

The ultimate goal of this project is the creation of a unified FPGA design, comprising both the feature extraction and neural network processing, capable of real-time automated classification. By real time we mean that the entire image is analysed in less than one second. Several additional steps are required to achieve this objective, namely further optimisation of the feature extraction stage. We have already elaborated several ideas, that combine both hardware and algorithm optimisation, outlined in Sect. 6.2.6. Combining these ideas together should, in principle, allow us to speed up the process by another order of magnitude. And since the FPGA resource utilisation for both stages is sufficiently low—that is, both stages can be combined into one design and run in parallel—the overall processing time will effectively be under a second. Such a design could then be used in practical clinical applications.

Let me conclude this chapter with another personal note. Besides the promising results presented above, this trip to Texas turned to be a very enriching experience. And I am not talking about moving to a far away continent for several months and living there alone for the first time—that is a totally different story, irrelevant to the present thesis. What matters here is that this internship was a great opportunity to start a new project, in a different field, alone and from scratch. In fact, while Dr. Milner supplied the main idea, all the following decisions were mine to take, including the choice of hardware, that was purchased following my instructions. The completion[5] of this project gives an important boost in confidence and inspiration, on top of valuable technical experience. All these factors will be important for my future postdoctoral research. And I am glad I could conclude my PhD with such an impactful journey.

## References

1. Lusis, Aldons J. 2000. Atherosclerosis. *Nature* 407 (6801): 233–241 (Sept 2000), issn: 0028-0836.
2. Kolodgie, Frank, D., Allen P. Burke, Andrew Farb, Herman K. Gold, Junying Yuan, Jagat Narula, Aloke V. Finn, and Renu Virmani. 2001. The thin-cap fibroatheroma: A type of vulnerable plaque: The major precursor lesion to acute coronary syndromes. *Current Opinion in Cardiology* 16 (5): 285–292.

---

[5]Technically, the project is not completed yet, but we achieved what we intended to during my stay in Texas.

3. Yabushita, Hiroshi, Brett E. Bouma, Stuart L. Houser, H. Thomas Aretz, Ik-Kyung Jang, Kelly H. Schlendorf, Christopher R. Kauffman, Milen Shishkov, Dong-Heon Kang, Elkan F. Halpern, and Guillermo J. Tearney. 2002. Characterization of human atherosclerosis by optical coherence Tomography. *Circulation* 106 (13): 1640–1645. issn: 0009-7322.
4. Bouma, Brett E., Martin Villiger, Kenichiro Otsuka, and Wang-Yuhl Oh. 2017. Intravascular optical coherence tomography (Invited). *Biomedical Optics Express* 8 (5): 2660–2686.
5. John, Noble, Harry Lemoine Greene and Fred F Ferri. 2001. Textbook of primary care medicine: Mosby.
6. Chamie, Daniel, Zhao Wang, Hiram Bezerra, Andrew M. Rollins and Marco A. Costa. 2011. Optical coherence tomography and fibrous cap characterization. *Current Cardiovascular Imaging Report* 4 (4): 276–283. 21949565[pmid]. issn: 1941-9066.
7. Wikipedia 2017. *Atherosclerosis.* http://en.wikipedia.org/wiki/Atherosclerosis.
8. Tavora, Fabio, Nathaniel Cresswell, Ling Li, David Fowler, and Allen Burke. 2010. Frequency of acute plaque ruptures and thin cap atheromas at sites of maximal stenosis. pt. *Arquivos Brasileiros de Cardiologia* 94: 153–159.
9. Garcia-Garcia, Hector M, Ik-Kyung Jang, Patrick W. Serruys, Jason C. Kovacic, Jagat Narula, and Zahi A. Fayad. 2014. Imaging plaques to predict and better manage patients with acute coronary events. *Circulation Research* 114 (12): 1904–1917.
10. Goode, Barbara G. Optical coherence tomography/cardiology: Totally tubular: Cardiovascular OCT goes prime time. In: *BioOptics World (2010).* http://www.biopticsworld.com/articlesprint/volume-3/issue-4/features/optical-coherence.html.VI.5. References 167
11. Jaguszewski, Milosz, and Ulf Landmesser. 2012. Optical coherence tomography imaging: Novel insights into the vascular response after coronary stent implantation. *Current Cardiovascular Imaging Reports* 5 (4): 231–238.
12. Baruah, Vikram L, Aydin Zahedivash, Taylor B. Hoyt, Deborah Vela, L. Maximilian Buja, Thomas E. Milner, and Marc D. Feldman. 2016. Abstract 19246: Histology-Validated neural networks enable accurateplaque tissue and thin-capped fibroatheroma characterization through intravascular optical coherence tomography. *Circulation* 134(Suppl 1), A19246-A19246. issn: 0009-7322.
13. Baruah, Vikram L., Aydin Zahedivash, Hoyt Taylor, Austin McElroy, Deborah Vela, L. M. Buja, Thomas Milner, and Marc Feldman. 2016. TCT-570 Histology-validated neural networks enable plaque tissue and thin-capped Fibroatheroma characterization through intravascular optical coherence tomography based virtual histology. English. *Journal of the American College of Cardiology* 68 (18).
14. Baruah, Vikram, Aydin Zahedivash, Taylor Hoyt, Austin McElroy, Deborah Vela, L. Maximilion Buja, Andrew Cabe, Meagan Oglesby, Arnold Estrada, Piotr Antonik, Thomas E. Milner, and Marc D. Feldman. Automated coronary plaque characterization using intravascular optical coherence tomography and a smart-algorithm approach – virtual histology OCT. *Journal of the American College of Cardiology: Cardiovascular Imaging.* (under review).
15. Haralick, Robert M., Karthikeyan Shanmugam, et al. 1973. Textural features for image classification. *IEEE Transactions on Systems, Man, and Cybernetics* 3 (6): 610–621.
16. Yang, Xiaofeng, Srini Tridandapani, Jonathan J. Beitler, David S. Yu, Emi J. Yoshida, Walter J. Curran, and Tian Liu. 2012. Ultrasound GLCM texture analysis of radiation-induced parotid-gland injury in head-and neck cancer radiotherapy: An in vivo study of late toxicity. *Medical Physics* 39 (9): 5732–5739.
17. Hall-Beyer, Mryka. *A grey level co-occurrence matrix tutorial,* Feb 2007. http://www.fp.ucalgary.ca/mhallbey/tutorial.htm.
18. IEEE standard VHDL language reference manual. *ANSI/IEEE Std 1076-1993* (1994).
19. Pedroni, Volnei, A. 2004. *Circuit design with VHDL.* MIT Press.
20. Math Works Benelux. 2017. *Hyperbolic tangent sigmoid transfer function.* http://nl.mathworks.com/help/nnet/ref/tansig.html.
21. Math Works Benelux. 2017. *Log-sigmoid transfer function.* http://nl.mathworks.com/help/nnet/ref/logsig.html.

# Chapter 7
# Conclusion and Perspectives

*Everything happened by chance, as planned.*

Life humour

The story of my Ph.D slowly but inevitably comes to an end. For almost four years I have been working in a fusion of machine learning, FPGA design, optics and electronics. There was absolutely no way to be bored with such a variety, and I could not be more happy about it. To be completely honest, I genuinely loved what I was doing and I feel very thankful towards the past me for making the right choices that brought me here. Speaking of the past, now is a good time to draw a short personal conclusion and compare what has been and what it had become. The whole picture looks very pleasing—so much has changed where it needed to, while the core values remained true. Some say that after graduating from a university in science, the best and practically the only skill one keeps is the ability to learn. For 5 years I have been practising and mastering this skill, and thoroughly applied it through the 4 years of the Ph.D. And I must recognise that ULB did a very good job, now that I can see how little I knew back then in 2013, and how comfortable I feel now in a field that used to look so scary…Not to be misunderstood, I most certainly realise that I have merely sampled the very tip of the iceberg, and there is still so much more to learn. But now that the method has been acquired, the only thing that remains is but to keep on rolling.

Enough personal comments for now, let us get back to the main topic and actually conclude this thesis. Prior to my work, experiments on analogue reservoir computing had mostly investigated the reservoir layer, i.e. different nonlinear nodes, multiplexing schemes or coding techniques (see e.g. [1, 2], or [3] for a more complete review). The optimisation of the readout weights and the computation of the output, on the other hand, was done using (somewhat rudimentary) offline training. One notable advance beyond these minimal systems was the implementation of analogue input

© Springer International Publishing AG, part of Springer Nature 2018     161
P. Antonik, *Application of FPGA to Real-Time Machine Learning*,
Springer Theses, https://doi.org/10.1007/978-3-319-91053-6_7

and output layers [4, 5]. However, the latter experiments failed to achieve state-of-the-art results due to incompatibility of offline training with complex and nonlinear readout layers. Recent trends in neural networks, together with several ideas coming from how the brain functions suggest that there are many additional ways in which a reservoir computer can be used, and in particular, different ways in which it can be trained. When I started my thesis the time was ripe to explore these more advanced applications of analogue reservoir computing. By interfacing a photonic reservoir computer with a FPGA we could address some of these challenges.

We performed three full experiments, with a fourth in progress. First, we trained the opto-electronic reservoir online (see Chap. 2). This provided shorter experimental runtimes and more accurate (and significantly lower) error rates than in previous experiments, and more importantly the online-trained system could equalise drifting and switching channels. Second, with a fast Gbit Ethernet interface between the FPGA and a PC, we could implement the error backpropagation algorithm in hardware (see Chap. 3). As a consequence we could significantly improve performance on three hard tasks in reasonable time. One of the fascinating aspects of this experiment is that it showed that a complex training method (error backpropagation) could at least partially be done in hardware. Third, exploiting the fast computational capability of the FPGA, we realised a RC with a output feedback and thereby demonstrated a system capable of solving completely new kinds of tasks, such as generating periodic and chaotic time series (see Chap. 4). Finally, combining the advantages of online training with the aspiration for an efficient analogue readout, we started designing an online-trained analogue readout layer and reported very promising preliminary numerical results (see Chap. 5). We chose to delay the development of the experiment to accommodate an additional and very interesting project (see Chap. 6).

These experiments open many new avenues of research, some short-term, which are rather simple extensions of the work already realised, others longer-term which would require much more investment.

*Advanced Online Learning.* Our first experiment with the FPGA board (see Chap. 2) showed that a photonic reservoir computer could be trained online, and that the idea worked quite well. Now that the feasibility has been demonstrated, one can think of improving the realisation. Most importantly, we used a basic and very slow training algorithm that requires several seconds to converge on a setup that runs on the microsecond timescale. As has been discussed in Chap. 2, Anteo Smerieri investigated two much more efficient algorithms: recursive least squares and reward-modulated Hebbian learning. We did not implement them back then because I was only starting with the FPGA designs and did not have the necessary skills. The task seems feasible now, and the time investment should be substantially rewarded. In short, the weights would converge 1000 times faster (roughly, $10^3$ updates instead of $10^6$). Several less crucial but still important improvements could also be added to the FPGA design, with a more efficient resource utilisation and a faster communication scheme between the computer and the FPGA.

In a much longer-term perspective, one could imagine a system in which the online training is done in hardware. That is, one could conceive an analogue system that implements the output layer (including multiplication by the readout weights), but also an analogue system that computes an error signal and uses it to optimise the readout weights following the update equation (such as Eq. 2.4).

*Fully-Hardware Backprop.* The backpropagation experiment by Michiel Hermans (see Chap. 3) demonstrated the advantage of training the input mask. While it is now more challenging to improve the actual RC performance, one can further increase the efficiency of the experiment. As explained in Chap. 3, most of the computations were performed on the computer. The FPGA chip should take care of that instead for a much faster setup. As a reminder, one of the limiting factors in our experiments was the slow convergence rate. This was especially problematic with the experiment on the TIMIT task, that took two weeks to complete. Optimising the input mask on the FPGA would resolve this bottleneck.

A longer term challenge is to figure out how to implement the error backpropagation algorithm in other systems than the one we used. And an even longer term challenge would be to implement it entirely in the analogue domain, without using any digital help.

*Output Feedback.* The story of the reservoir computer with output feedback is a different one. We spent a significant amount of time on this experiment but could not find a simple way of improving the results due to the noise floor of our experimental system. This, however, does not mean that the situation is hopeless. There are several ideas we could try. The most conservative one is to keep the opto-electronic reservoir scheme, but build an entirely new experimental setup with low-noise components. This requires in-depth knowledge of the electronics and optics market and a significant financial investment. A more liberal idea is to get rid of the opto-electronic reservoir and switch to a different experiment. My fellow Ph.D colleague, Quentin Vinckier, developed a new optical reservoir computer with a passive cavity [2], with very low level of noise, since there are now active components inside the delay loop. The challenge here is first to master the anything-but-trivial experiment, much more complex than the opto-electronic reservoir, and only then to build a new interface for the FPGA. Several other ideas have been proposed in our paper [6], such as the use of conceptors [7, 8], but we did not have the time to implement them.

Another challenge would be to realise a fully analogue reservoir computer with output feedback. For instance, one could consider using the analogue output presented in [5], and feed this back into the reservoir. Implementing this is a major challenge.

*Online-Trained Analogue Readout.* The online training of an analogue readout layer is probably the most promising short-term goal. In fact, we have paved most of the road. Therefore, the plan of action is very simple here: just do it, now experimentally.

*Ultra-Fast Systems.* Another challenge would be to implement the same experiments I have reported in my thesis, but using much faster systems. Experimental reservoir computers based on delay dynamical systems can be much faster than the experiment I implemented, see e.g. [9, 10]. Can online training, physical error backpropagation or output feedback be implemented using these reservoirs? A major challenge here is that FPGAs themselves may no longer be able to follow the speed of the reservoir.

Besides the experiments discussed above, during the last year of my thesis, I took an amazing opportunity to spend five months in the sunny Texas, applying my FPGA skills and machine learning knowledge to the field of coronary imaging. Exploiting the parallel computational capacities of the FPGA, we could reduce the runtime of the automated plaque characterisation algorithm, based on artificial neural networks and developed at UTexas prior to my arrival, from days to tens of seconds (see Chap. 6). This speedup brings the method much closer to potential clinical use. Future investigations, with a possible collaboration between ULB and UTexas, may see the neural network replaced by an optical reservoir computer, tightly linked to the OCT device used for artery imaging. This also opens several directions for future research.

*Real-Time Plaque Characterisation on FPGA.* The continuation of my project with UTexas seems a very exciting research direction. That is, the two main design blocks have been developed, and all that remains is to connect them together, with some hardware optimisation. However, there are situations where one should take things slowly, and this is one of them. Dr. Milner's team had already came to realisation that, while the neural network classifier with feature extraction is a functioning approach, it may not be the simplest or the most optimal one. A few quick tests on my own, not presented in this thesis, have suggested that the neural network could be simplified, and that not all of the 300 features are necessary. The details are yet to be confirmed, but the general idea sounds promising. Which brings us to the next point.

*Efficient Plaque Classification Algorithm.* The neural network classifier for plaque characterisation allowed to achieve state-of-the-art results in terms of sensitivity and specificity. No one could possibly deny that it is a good classifier. But it may not be the best one. After some analysis of the network structure and its core characteristics, it should be possible to understand what exactly makes it a good classifier and how complex it should be. Again, this is an investigation that we have already started in Austin, but it is too soon to draw any conclusions.

*Novel Applications of RC.* During the summer of 2016, I supervised an intern who was working on applications of reservoir computing to various pattern recognition tasks. Although we obtained some promising preliminary results, the idea requires to be further investigated. The motivation arises from my personal point of view. While reservoir computing is a very elegant model, it seems to me that its optical implementations are being investigated as a "toy", and they are yet to find a "killer" application. This is why, I believe, finding the perfect problem to solve with optical reservoir computing is an interesting avenue of research. And it turns out that my internship in Texas might just have brought a candidate for such an application. Consider the following scenario. On the one hand, there is the optical coherence tomography for scanning the arteries. On the other—a neural network classifier in charge of analysing the OCT images and characterising different tissue types. Would it be possible to replace the neural network with an optical reservoir computer and fit it nicely alongside OCT? Such an application would exploit all the main benefits of photonic RC—high-speed optical processing, low energy consumption, ease of training the reservoir—and most certainly attract the key players in the $500 million coronary imaging market [11].

In conclusion, there are many directions in which the results presented in the present thesis could be extended. Some are relatively simple and easy to follow, others are more ambitious and challenging. But the key point is the following. When I joined OPERA-Photonique four years ago, photonic reservoir computing had been demonstrated in its simplest versions, and the team had an FPGA board and an idea of what to do with it. Four years later, several advanced information processing tasks have been demonstrated by interfacing the FPGA with the experiment, opening up many perspectives for future research. Thus today the lab still possesses the same FPGA, but now with half-a-dozen new ideas. Therefore, while it is time for me to bow out, the story of FPGA-enhanced photonic reservoir computing is only beginning!

On a final note, let us take another quick peek into the future. I have presented several posters at machine learning conferences, with audiences expert in neural networks, but relatively new to optical implementations of reservoir computing. Quite often, participants were curious about the future and practical applications of our experiments. In particular, at my last Benelearn conference in June 2017, a dutch scholar, fascinated by our work, asked whether optical reservoir computing would be "the future". While I definitely admire the simplicity of the concept and the processing speeds that can be achieved with fast optical setups, I think that this is an overstatement. So what could we actually expect from optical RC in the near future?

As Michiel Hermans often told us, the computational power of a machine learning algorithm depends on the number of trainable parameters. In reservoir computing, this number is significantly reduced, which favours the simplicity of the system but deteriorates its performance on complex tasks. For comparison, experimental reservoir computers typically count at most a few hundred of trainable parameters, while deep learning, that currently dominates the machine learning world, totals millions of them. Therefore, I do not expect reservoir computers to become ubiquitous—they are too simple to deal with challenging tasks, such as image processing. But I believe there is a niche where reservoir computing will prevail over the competition. A niche that requires, for instance, moderately complex processing of optical signals at very high speeds, with low energy consumption. In that case, all of the main advantages of optical RC could be exploited.

A realistic commercial application of optical RC does not seem possible without a fast and robust device, probably integrated on a chip. This endeavour would probably require joining the efforts of several research group across Europe: the integrated reservoir [12] from UGent, the high-speed systems [10, 88] from IFISC and FEMTO-ST, and the low-energy passive coherent cavity [2] from our group. It is virtually impossible to tell how much time such a large-scale collaboration would take, but the resulting system could attract multiple clients from the industry. For instance, it could be used in optical communications for routing the packets without converting them to digital domain, thus increasing processing speed and cutting costs on analogue-to-digital converters. Furthermore, as my internship in Texas has recently suggested, medical imaging, such as intravascular OCT, could also benefit from embedded optical computing for image pre-processing or analysis. To conclude, I do not think that optical reservoir computing is *the* future, but I believe it will most certainly be part of it.

# References

1. Duport, François, Bendix Schneider, Anteo Smerieri, Marc Haelterman, and Serge Massar. 2012. All-optical reservoircomputing. *Optics Express* 20: 22783–22795.
2. Vinckier, Quentin, François Duport, Anteo Smerieri, Kristof Vandoorne, Peter Bienstman, Marc Haelterman, and Serge Massar. High-performancephotonic reservoir computer basedon a coherently driven passive cavity. *Optica* 2.5: 438–446.
3. Van der Sande, Guy, Daniel Brunner, and Miguel C. Soriano. 2017. Advancesin photonic reservoir computing. *Nanophotonics* 6 (3): 561–576.
4. Smerieri, Anteo, François Duport, Yvan Paquot, Benjamin Schrauwen, Marc Haelterman, and Serge Massar. Analog readout for optical reservoir computers. *Advances in Neural Information Processing Systems*, 944–952.
5. Duport, François, Anteo Smerieri, Marc Haelterman AkramAkrout, and Serge Massar. 2016. Fully analogue photonic reservoir computer. *Scientific Reports* 6: 22381.
6. Antonik, Piotr, Marc Haelterman, and Serge Massar. 2017. Brain-inspired photonic signal processor for generating periodic patterns and emulatingchaotic systems. *Physical Review Applied* 7: 054014.
7. Jaeger, Herbert. 2014. Conceptors: An easy introduction. In: CoRR abs/1406.2671.
8. Jaeger, Herbert. 2014. Controlling recurrent neural networks by conceptors. In: CoRR abs/1403.3369.
9. Brunner, Daniel, Miguel C. Soriano, Claudio R. Mirasso, and Ingo Fischer. 2013. Parallel photonic informationprocessing at gigabyte per second data rates using transient states. *Nature Communications* 4: 1364.
10. Larger, Laurent, Antonio Baylón-Fuentes, Romain Martinenghi, Vladimir S. Udaltsov, Yanne K. Chembo, and Maxime Jacquot. 2017. High-speed photonic reservoir computing using a time-delay-based architecture: Million words per second classification. *Physical Review X* 7: 011015.
11. Goode, Barbara G. 2010. Optical coherence tomography/cardiology: totallytubular: Cardiovascular OCT goes primetime. 2010. *BioOptics World*. http://www.bioopticsworld.com/articles/print/volume-3/issue-4/features/optical-coherence.html.
12. Vandoorne, Kristof, Pauline Mechet, Thomas Van Vaerenbergh, Martin Fiers, Geert Morthier, David Verstraeten, Benjamin Schrauwen, Joni Dambre, and Peter Bienstman. 2014. Experimental demonstration of reservoir computing on a silicon photonics chip. *Nature Communications* 5: 3541.

# Author's Curriculum Vitae

**Contact information**:
LMOPS EA-4423
CentraleSupélec, Metz campus
2 rue Edouard Belin
F-57070 Metz, France
+33 (0)3 87 76 47 96
piotr.antonik@supelec.fr

**Personal information**:
Age: 28 years (1989)
Birthplace: Minsk, Belarus
Citizenship: Belgian
Residence: Metz, France
+33 (0)6 43 80 21 22
peter.antonik@gmail.com

**Profile**

- Five years of fruitful research in highly interdisciplinary domain
- Active collaborations with national and international partners
- Design and development of advanced optical and opto-electronic experimental systems
- Mastery of FPGA designing (Xilinx platforms in particular)
- High expertise in artificial neural networks, general knowledge of machine learning
- Rich experience in teaching (lab assistant) and intern supervision
- Skilled at programming, oral presentations and scientific writing
- Successful in autonomous and team work
- Bilingual (Russian and French), fluent in written and spoken English

© Springer International Publishing AG, part of Springer Nature 2018
P. Antonik, *Application of FPGA to Real-Time Machine Learning*,
Springer Theses, https://doi.org/10.1007/978-3-319-91053-6

**Education**

- Ph.D. in Physics, Université libre de Bruxelles, Brussels (Oct. 2013–Sep. 2017)
- M.S. in Physics (with great honours), Université libre de Bruxelles, Brussels (Sep. 2011–Sep. 2013)
- B.S. in Physics (with great honours), Université libre de Bruxelles, Brussels (Sep. 2008–Aug. 2011)

**Research Positions**

**October 2017—present**:    Postdoctoral position at CentraleSupélec

- Position: LMOPS EA-4423 Lab, CentraleSupélec, Metz campus, Metz, France
- Research interests: optical neuromorphic computing, analogue time-delay reservoir computing, large-scale free-space optical reservoir computing, wide-band opto-electronic oscillators for advanced signal processing

**January—May 2017**:    Internship at UTexas

- Position: Department of Biomedical Engineering, University of Texas at Austin, Austin, USA
- Topic: Application of artificial neural networks to real-time medical image processing and diagnosis

**October 2013—September 2017**:    PhD Student at ULB

- Position: Laboratoire d'Information Quantique and Service OPERA-Photonique, Université libre de Bruxelles, Brussels, Belgium
- Topic: Application of FPGA to real-time machine learning: hardware reservoir computers and software image processing

**September 2012—May 2013**:    Master's Project at ULB

- Position: Laboratoire d'Information Quantique and Service OPERA-Photonique, Université libre de Bruxelles, Brussels, Belgium
- Topic: Reservoir computing with delayed-feedback laser

**Publications**

**Published**:

- 5 peer-reviewed journal articles (including 1 PRL, 1 IEEE TNNLS, 1 Phys. Rev. Appl.)
- 8 full-length conference papers (including 1 IJCNN, 2 ICONIPs),
- 4 conference abstracts (including 2 AAAIs).
  Aggregate impact factor: 24.438 (2016). Total citations: 48 (18.03.2018).

**In progress**:    3 peer-reviewed journal articles, 1 full-length conference paper, 3 conference abstracts.

See full publications list below.

**Teaching**

**2016–2017**:  Lab assistant: general physics (first year students, 28 hours), optics (second year students, 40 hours)

**2015–2016**:  Lab assistant: general physics (first year students, 28 hours), optics (second year students, 41 hours), electronics (second year students, 36 hours)

**2014–2015**:  Lab assistant: general physics (first year students, 28 hours), electronics (second year students, 36 hours)

Total teaching: 237 hours in 3 years.

**Grants, Fellowships and Awards**

- Springer Theses Award (2018)
- F.R.S.-FNRS Research Grant J.0040.16. Photonic Reservoir Computing (2016–2017)
- F.R.S.-FNRS Research Fellow (Oct. 2013–Sep. 2017)
- Honourable Mention at the 38th International Physics Olympiad, Vietnam, 2008
- Winner of the Belgian Physics Olympiad, 2007

**Activities**

- Reviewer for IEEE TNNLS, Neurocomputing, Optics Express and Cognitive Computation journals.

# Author's Publications

**Journal Papers**

(1) Piotr Antonik, François Duport, Michiel Hermans, Anteo Smerieri, Marc Haelterman, and Serge Massar. "Online Training of an Opto-Electronic Reservoir Computer Applied to Real-Time Channel Equalization". *IEEE Transactions on Neural Networks and Learning Systems* 28:11 (2017), pp. 26862698.

- Impact factor: 6.108 (2016).

(2) Michiel Hermans, Piotr Antonik, Marc Haelterman, and Serge Massar. "Embodiment of Learning in Electro-Optical Signal Processors". *Phys. Rev. Lett.* 117 (12 2016), p. 128301.

- Impact factor: 8.462 (2016).

(3) Piotr Antonik, Michiel Hermans, Marc Haelterman, and Serge Massar. "Random Pattern and Frequency Generation Using a Photonic Reservoir Computer with Output Feedback". *Neural Processing Letters* (2017), pp. 1–14.

- Impact factor: 1.620 (2016).

(4) Piotr Antonik, Marc Haelterman, and Serge Massar. "Online Training for High-Performance Analogue Readout Layers in Photonic Reservoir Computers". *Cognitive Computation* 9 (2017), pp. 297–306.

- Impact factor: 3.44 (2016).

(5) Piotr Antonik, Marc Haelterman, and Serge Massar. "Brain-Inspired Photonic Signal Processor for Generating Periodic Patterns and Emulating Chaotic Systems". *Phys. Rev. Applied* 7 (2017), p. 054014.

- Impact factor: 4.808 (2016).

**In Progress**

(1) Piotr Antonik, Marvyn Gulina, Jaël Pauwels, and Serge Massar, "Using a reservoir computer to learn chaotic attractors, with applications to chaos synchronisation and cryptography," *Phys. Rev. E*, under review.
(2) Vikram Baruah, Aydin Zahedivash, Taylor Hoyt, Austin McElroy, Deborah Vela, L. Maximilian Buja, Andrew Cabe, Meagan Oglesby, Arnold Estrada, Piotr Antonik, Thomas E. Milner, Marc D. Feldman, "Automated Coronary Plaque Characterization with Intravascular Optical Coherence Tomography and a Smart-Algorithm Approach—Virtual Histology OCT," *JACC: Cardiovascular Imaging*, under review.
(3) Piotr Antonik, Vikram Baruah, Aydin Zahedivash, Austin McElroy, Taylor Hoyt, Serge Massar, Marc Feldman, and Thomas Milner, "FPGA processing of IV-OCT scans for high-speed automated coronary plaque characterisation," *IEEE Transactions on Biomedical Engineering*, in preparation.

**Conference Papers**

(1) Piotr Antonik, Anteo Smerieri, François Duport, Marc Haelterman, and Serge Massar. "FPGA implementation of reservoir computing with online learning". In: *24th Belgian-Dutch Conference on Machine Learning.*
(2) Piotr Antonik, François Duport, Anteo Smerieri, Michiel Hermans, Marc Haelterman, and Serge Massar. "Online training of an opto-electronic reservoir computer". In: *APNNAs 22th International Conference on Neural Information Processing.* Vol. 9490. LNCS. 2015, pp. 233–240.
(3) Piotr Antonik, François Duport, Anteo Smerieri, Michiel Hermans, Marc Haelterman, and Serge Massar. "Improving performance of opto-electronic reservoir computers with online learning". In: *20th Annual Symposium of the IEEE Photonics Society Benelux Chapter.* 2015.
(4) Piotr Antonik, Michiel Hermans, François Duport, Marc Haelterman, and Serge Massar. "Towards pattern generation and chaotic series prediction with photonic reservoir computers". In: *SPIEs 2016 Laser Technology and Industrial Laser Conference.* Vol. 9732. 2016, 97320B.
(5) Piotr Antonik, Michiel Hermans, Marc Haelterman, and Serge Massar. "Towards adjustable signal generation with photonic reservoir computers". In: *25th International Conference on Artificial Neural Networks.* Vol. 9886. 2016.

(6) Piotr Antonik, Michiel Hermans, Marc Haelterman, and Serge Massar. "Pattern and frequency generation using an opto-electronic reservoir computer with output feedback". In: *APNNSs 23th International Conference on Neural Information Processing*. Vol. 9948. LNCS. 2016, pp. 318–325.

(7) Akram Akrout, Piotr Antonik, Marc Haelterman, and Serge Massar. "Towards autonomous photonic reservoir computer based on frequency parallelism of neurons". In: *Proc. SPIE*. Vol. 10089. 2017, 100890S-100890S-7.

(8) Piotr Antonik, Michiel Hermans, Marc Haelterman, and Serge Massar. "Photonic Reservoir Computer With Output Feedback for Chaotic Time Series Prediction". In: *2017 International Joint Conference on Neural Networks*. 2017.

**In Progress**

(1) Piotr Antonik, Marvyn Gulina, Jaël Pauwels, Damien Rontani, Marc Haelterman, and Serge Massar, "Spying on chaos-based cryptosystems with reservoir computing," in *2018 International Joint Conference on Neural Networks*, accepted, to appear.

**Conference Abstracts**

(1) Piotr Antonik, Marc Haelterman, and Serge Massar. "Improving Performance of Analogue Readout Layers for Photonic Reservoir Computers with Online Learning". In: *AAAI Conference on Artificial Intelligence*. 2017.

(2) Piotr Antonik, Michiel Hermans, Marc Haelterman, and Serge Massar. "Chaotic Time Series Prediction Using a Photonic Reservoir Computer with Output Feedback". In: *AAAI Conference on Artificial Intelligence*. 2017.

(3) Piotr Antonik, Marc Haelterman, and Serge Massar. "Predicting chaotic time series using a photonic reservoir computer with output feedback". In: *26th Belgian-Dutch Conference on Machine Learning*. 2017.

(4) Piotr Antonik, Marc Haelterman, and Serge Massar. "Towards high-performance analogue readout layers for photonic reservoir computers". In: *26th Belgian-Dutch Conference on Machine Learning*. 2017.

**In Progress**

(1) Marvyn Gulina, Piotr Antonik, Jael Pauwels, Damien Rontani, Marc Haelterman, and Serge Massar, "Cracking chaos-based cryptography with reservoir computing," in *2018 International Symposium on Nonlinear Theory and Its Applications*, to appear.

(2) Piotr Antonik, Damien Rontani, Marc Haelterman, and Serge Massar, "Towards online-trained analogue readout layer for photonic reservoir computers," in *2018 International Symposium on Nonlinear Theory and Its Applications*, to appear.

(3) Piotr Antonik, Nicolas Marsal, Daniel Brunner, and Damien Rontani, "Performance of large photonics networks for reservoir computing," in *2018 International Symposium on Nonlinear Theory and Its Applications*, to appear.

Printed in the United States
By Bookmasters